Modern Electrical Engineering

Modern Electrical Engineering

(Objected-Oriented Approach)

Ram Bansal 'Vigyacharya'

CBS PUBLISHERS & DISTRIBUTORS PVT. LTD.
New Delhi • Bengaluru • Chennai • Kochi • Mumbai • Pune

ISBN: 81-239-0618-8

First Edition: 1998
Reprint: 2001, 2008, 2010, 2015

Published by:
Satish Kumar Jain for CBS Publishers & Distributors Pvt. Ltd.,
CBS Plaza, 4819/XI Prahlad Street, 24 Ansari Road,
Daryaganj, New Delhi – 110002, India
delhi@cbspd.com, cbspubs@airtelmail.in • www.cbspd.com
Ph.: 23289259, 23266861, 23266867 • Fax: 011-23243014

Branches:
- ***Bengaluru:*** Seema House, 2975, 17th Cross, K.R. Road,
  Bansankari 2nd Stage, Bengaluru - 560070
  Ph: +91-80-26771678/79 • Fax: +91-80-26771680
  E-mail: cbsbng@gmail.com, bangalore@cbspd.com
- ***Chennai:*** No. 7, Subbaraya Street, Shenoy Nagar, Chennai - 600030
  Ph: +91-44-26681266, 26680620 • Fax: +91-44-42032115
  E-mail: chennai@cbspd.com
- ***Kochi:*** 36/14, Kalluvilakam, Lissie Hospital Road, Kochi - 682018
  Ph: +91-484-4059061-65 • Fax: +91-484-4059065
  E-mail: cochin@cbspd.com
- ***Mumbai:*** 83-C, Dr. E. Moses Road, Worli, Mumbai - 400018
  Ph: +91-9833017933, 022-24902340/41 • E-mail: mumbai@cbspd.com
- ***Pune:*** Bhuruk Prestige, Sr. No. 52/12/2+1+3/2,
  Narhe, Haveli (Near Katraj-Dehu Road Bypass), Pune - 411041
  Ph: +91-20-64704058/59, 32342277 • E-mail: pune@cbspd.com

Printed at :
J.S. Offset Printers, Delhi

Dedicted to

SHRI SATYA PAL and SHRIMATI SHAKUNTALA GARG

my sources of inspiration and moral support

Dedicated to

SHRI SATYA PAL and SHRIMATI SHAKUNTALA GARG

my sources of inspiration and moral support

About the Book

Electrical Engineering has undergone a sea-change in last few years under the impact of Electronics and Computers, these forming essential parts of electrical engineering curriculum in education, services and practice. This book attempts to anchor the reader within the depths of the subject through objective-type advanced-level questions and the abstracts of the chapters. No doubt, readers will appreciate the limitations on the author to provide a view of such a vast subject in such a few pages. The Abstracts of the chapters are a mere reminder to the reader and not the study material. The topics have been seggregated and presented with a logic in mind. These include :

- Basic Principles
- Materials and Properties
- Basic Electronics
- Control Engineering
- Power Systems
- Electrical Machines
- Electric Drives and Traction
- Measurements and Instrumentation
- Electromagnetic Waves and Communication
- Power Electronics
- Digital Electronics and Computers
- Electrothermics and Electroluminance.

Engineering is a social science, hence the stress is on the applicability of the technological developments with blending of theory and practice.

About the Author

Ram Bansal 'Vigyacharya' served various giants to small industries for a decade, then he worked as an Industrial Consultant in project engineering, and modernization services for 12 years. Now the aims to recreate, what he is, to leave the imprints for the future engineering generations. He studies and presently whole of his time is devoted to the studies and writings on Electrical Engineering, Computers and Spiritual Sciences. He had his prime education at University of Roorkee and is on the guest faculty of Malviya Regional Engineering College, Jaipur.

Preface

A book is not merely a pack of papers but a living and vibrant channel of communication between the author and the reader. With this view in the mind, I started writing this book but due to paucity of space in a book of this type, the above concept could not be fully implemented. The whole subject of electrical engineering, which the book is supposed to cover needs, about ten thousand pages of a fully-loaded text. Hence in the abstract portions of this book only a few hints of the subject could be provided. However, the questionaire part covers advanced questions which cannot be answered on the basis of text in this book. For answering the questions, a reader must have a prior study of the subject from other text books, then only the hints provided in the book can be of some use, as a reminder of the subject.

The book is aimed at testing the student with an objective of preparing him for the competitive examinations such as IAS, IES, Post-graduate admission test, AMIE and various departmental examinations held for various specific purposes such as recruitment, promotions etc.

An effort has been made to limit the volume of the book by including only advanced level questions, as simpler questions have been felt to be redundant in a book of this nature. Modern thought has been, in all aspect of life in general and engineering in particular, to update the output through feed-back, without which the output is likely to deviate from the desired characteristics. Although a lot of interaction has been placed during the preparation of the manuscript, a lot more is wanted. The readers of all type are requested to provide their valuable suggestions for the improvement of the future editions of the book.

A few gentlemen have been, in one way or the other, instrumental in writing this book. Dr. Mool Singh, Head of Electrical Engineering Department, Malviya Regional Engineering College, Jaipur is the first and the foremost by bringing me in close contact with the engineering academy. Prof. R.S. Nirjar, Principal of the College, and Dr. S.P. Gupta, President Library have been kind enough in granting me the library facility without which the work would not have been of this standard. Mr. Deep Singh,

Assistant Librarian and many of his staff have always been of great and liberal help to me. All these gentlemen, not only deserve appreciation for their generosity but also my sincerest gratefulness.

Shri Man Mohan Singh Shekhawat, Assistant Professor, Agronomy, Agricultural Research Farm, Rajasthan Agricultural University, Jaipur and his family have been a source of moral and material support, without my asking for it and without any expectation of any return by them. In these difficult and competitive times, generosity of this couple is a rare pheromenon for which I would like to record my appreciation.

Jaipur **Ram Bansal 'Vigyacharya'**

Contents

1

Basic Principles

Q.1. Volt is basically equivalent to -

(a) joules per second, (b) joules per coulomb,
(c) coulombs per second, (d) none of the above.

Q.2 Ampere is basically equivalent to -

(a) joules per second, (b) joules per coulomb,
(c) coulombs per second, (d) none of the above.

Q.3 A dc generator is supplying current to a load. The direction of current inside the generator is -

(a) from positive to negative, (b) from negative to positive,
(c) alternating, (d) none of the above.

Q.4 Ohm's law is applicable to -

(a) only linear resistors,
(b) both linear and nonlinear resistors,
(c) none of the above.

Q.5 Which one of the following is false -

(a) 1 pico ohm = 10^{-21} giga ohm,
(b) 10^{15} nano ampere = 1 mega ampere,
(c) 1 mega ohm = 10^{-6} ohm,
(d) none of the above.

Q.6 Minimum-load condition of a generator means, across its terminals is -

(a) highest conductance, (b) lowest resistance,
(c) highest resistance, (d) none of the above.

Q.7 If we connect two lamps of 100 W, 100 V and 200 W, 200 V each, connected in series, to a 300 volt supply, what is expected to happen ?

(a) lamps will glow normally, (b) 100 W lamp will get fused,

(c) 200 W lamp will get fused, (d) none of the above.

Q.8 Watt is basically equivalent to -

(a) joules per second, (b) joules per coulomb,
(c) coulombs per second, (d) none of the above.

Q.9 All circuits can be drawn on a paper, without any overcrosses by the elements or branches of the circuit.

(a) true, (b) false,

Q.10 Inductors and capacitors never receive any power from the source.

(a) true, (b) false,
(c) partially true, (d) nonsense.

Q.11 Power received by an ideal inductor or capacitor, from a source, is -

(a) dissipated in the ambience,
(b) dissipated as heat,
(c) returned back to the source itself,
(d) none of the above..

Q.12 Effective value of an variable current is -

(a) average current,
(b) current/2,
(c) dc current required to deliver the same power,
(d) sinusoidal current required to deliver the same power.

Q.13 For a sinusoidal current wave, the effective value is -

(a) peak-value/2, (b) peak-value/3,
(c) average value, (d) none of the above.

Q.14 In India, the nominal single-phase domestic power supply peak voltage is -

(a) 220 V, (b) 230 V,
(c) 400 V, (d) none of the above.

Q.15 Usually, rms value of ac current is treated as effective value because -

(a) the usual supply is sinusoidal,
(b) it produces the same voltage difference,
(c) it produces the same power flow,
(d) none of the above.

Q.16 Usually, the rms value of current is assumed to be peak-value/2, because -

(a) the usual supply is sinusoidal,
(b) it produces the same voltage difference,

(c) it produces the same power flow,
(d) none of the above.

Q.17 If mean value of a function is zero, its rms value will be -
(a) zero, (b) peak-value/$\sqrt{2}$,
(c) average-value × 1.1, (d) none of the above.

Q.18 For a sinusoidal wave, the form factor is 1.1. The form factor being -
(a) rms value/average value, for full cycle,
(b) peak value/rms value for full cycle,
(c) peak value/average value for full cycle,
(d) none of the above.

Q.19 If V and I are the peak values of voltage and current respectively, of an ac supply, the average power consumption will be -
(a) VI x power-factor, (b) VI,
(c) ½ VI x power-factor, (d) none of the above.

Q.20 For a circuit, having inductive and capacitive elements, if the imaginary part of its impedance is zero, the circuit is said to be-
(a) resistive, (b) of real power,
(c) resonant, (d) none of the above.

Q.21 The frequency at which the inductive and capacitive effects of a circuit elements are neutralized by each other, is called -
(a) critical frequency, (b) cutoff frequency,
(c) resonance frequency, (d) none of the above.

Q.22 If L be the inductance and C the capacitance of a RLC circuit, the resonance frequency, in radians, is given by -
(a) $\sqrt{(LC)}$, (b) $1/\sqrt{(LC)}$,
(c) $\sqrt{(L/C)}$, (d) none of the above.

Q.23 Unit of magnetic permeability is -
(a) Farad per meter, (b) Henry per meter,
(c) Henry per second, (d) none of the above.

Q.24 Value of magnetic permeability, in Henrys per meter, of free space is -
(a) $4\pi \times 10^{-7}$, (b) $1/[4\pi \times 10^{-7}]$,
(c) $4\pi \times 10^{7}$, (d) none of the above.

Q.25 For a coil, the product of current flowing in it and its number of turns, is called -
(a) Electromotive force, (b) Magnetomotive force,
(c) Magnetic strength, (d) Magnetic intensity.

Q.26 For obtaining magnetic reluctance of a path, the MMF is to be devided by -

(a) length of the path, (b) flux density,
(c) area of the path, (d) none of the above.

Q.27 Inductance of a coil is proportional to -

(a) number of turns, (b) square of number of turns,
(c) inverse of number of turns, (d) none of the above.

Q.28 In circuit analysis, if two points are found to be at the same potential, circuit element currents and potentials will remain unchanged if these two points are -

(a) short-circuited,
(b) open-circuited,
(c) short-circuited or open-circuited,
(d) none of the above.

Q.29 A current source with a parallel internal resistance of R_s, wii provide maximum power to a resistive load, when the load resistance is equal to -

(a) zero, (b) infinite,
(c) Rs., (d) none of the above.

Q.30 A voltage source with a sries internal resistance of R_s, will deliver maximum power when the load resistance is equal to -

(a) zero, (b) infinite,
(c) Rs, (d) none of the above.

Q.31 An electrical circuit is considered to be a lossless circuit if its resistance is -

(a) zero only, (b) infinite only,
(c) zero or infinite, . (d) none of the above..

Q.32 In phasor mathematics, Euler's identity $V.e^{j\omega t + \theta}$) is equivalent to -

(a) $V \sin(\omega t + \theta) + j V \cos(\omega t + \theta)$,
(b) $V \cos(\omega t + \theta) + j V \sin(\omega t + \theta)$,
(c) $V \cos(\omega t - \theta) + j V \sin(\omega t - \theta)$,
(d) none of the above.

Q.33 In electrical circuits, term 'immitance' is used for -

(a) oppositeto resistance,
(b) opposite to impedance,
(c) jointly for impedance and admittance,
(d) none of the above.

Q.34 In electrical circuits, 'Siemens' is the unit of -

(a) admittance, conductance and susceptance,

(b) magnetic permeance,
(c) magnetic susceptibility,
(d) none of the above.

Q.35 In electrical circuits, susceptance is -
(a) inverse of reactance, (b) imaginary part of admittance,
(c) imaginary part of impedance, (d) none of the above.

Q.36 Superposition theorem of electrical circuit analysis is applicable for calculating power in (i) d.c. circuits, (ii) a.c. circuits fed with same frequency current sources, (iii) a.c. circuits fed with different frequency current sources.
(a) all the above are true, (b) only last two are true,
(c) only last one is true, (d) none of the above is true.

Q.37 In electrical circuits, effective value of a periodic current is equivalent to a dc current that delivers the same -
(a) average power, (b) average volt-drop,
(c) average energy, (d) none of the above.

Q.38 In electrical circuits, power factor is the ratio of -
(a) average power to peak power,
(b) average energy to peak energy,
(c) average power to apperent power,
(d) none of the above.

Q.39 In electric circuits, a lagging load means (i) purely inductive load, (ii) which has a volt-drop lagging the current, (iii) whose power factor is negative, (iv) whose power factor is positive.
(a) only (iv) is false, (b) only (iii) is false,
(c) only (ii) and (iv) are false, (d) all the above are false..

Q.40 In electric circuits, reactive power means -
(a) apperent power, (b) imaginary power,
(c) effective power, (d) none of the above.

Q.41 In electric circuits, a two-phase balanced system is that in which the two voltage phasors (i) are 180 degrees apart, (ii) sum to zero, (iii) equal and opposite to each other.
(a) all the above are true, (b) only (i) and (ii) are true,
(c) only (i) is true, (d) none of the above is true..

Q.42 Admittance parameters of a two-port network are obtained by -
(a) open-circuiting input/output terminals,
(b) short-circuiting input/output terminals,
(c) open-circuiting input and short-circuiting output terminals,
(d) none of the above.

Q.43 In a passive bilateral network, if a voltage V across branch A produces a current I in branch B, then the same voltage V put across branch B will produce a current in branch A, that will be equal to I. This is -

(a) Norton's theorem, (b) Reciprocity theorem,
(c) Thevnin's theorem, (d) none of the above.

Q.44 In a two-port bilateral, for the admittance parameters, the followingg is true -

(a) $Y_{11} = Y_{22}$, (b) $Y_{12} = Y_{21}$,
(c) $Y_{11} = Y_{12}$, (d) none of the above.

Q.45 If two two-port networks are connected in parallel, to get the admittance parameters of the combined network -

(a) all admittance parameters can be respectively added,
(b) only transfer admittance parameters can be respectively added,
(c) only input/output admittance parameters can be respectively added,
(d) none of the above is true.

Q.46 Impedance parameters of a two-port network are obtained by -

(a) open-circuiting input/output terminals,
(b) short-circuiting input/output terminals,
(c) open-circuiting input and short-circuiting output terminals,
(d) none of the above.

Q.47 Impedance parameters of two two-port networks can be respectively added if the two networks -

(a) are connected in series -one after the other,
(b) inputs and outputs are respectively series-connected,
(c) are connected in parallel,
(d) none of the above.

Q.48 The hybrid parameters of a two-port network, have the unit -

(a) siemens, (b) ohms,
(c) nepers, (d) none of the above..

Q.49 If V_1, V_2 are the input output voltages and I_1, I_2 the input output currents of a two-port network, two of the hybrid parameters are so related -

(a) $V_1 = h_{11}.I_1 + h_{12}.V_2$, (b) $V_1 = h_{11}.I_1 + h_{12}.I_2$,
(c) $I_1 = h_{11}.V_1 + h_{12}.V_2$, (d) none of the above.

Q.50 For a bilateral two-port network, the two of the hybrid parameters are so related -

(a) $h_{12} = h_{21}$, (b) $h_{12} = - h_{21}$,
(c) $h_{11} = h_{22}$, (d) none of the above.

Q.51 Hybrid parameters of two two-port networks can be respectively added if networks -
(a) are connected in cascade,
(b) are connected in series,
(c) are connected in parallel,
(d) inputs are connected in series and outputs in parallel..

Q.52 The transmission parameters of a two-port network are so related -
(a) $V_1 = A.V_2 - B.I_2$ and $I_1 = C.V_2 - D.I_2$,
(b) $V_1 = A.I_1 - B.I_2$ and $V_2 = C.I_1 - D.I_2$,
(c) $I_1 = A.V_1 - B.V_2$ and $I_2 = C.V_1 - D.V_2$,
(d) none of the above.

Q.53 If the two two-port networks are connected in cascade - one after the other, the transmission parameters of the combined network can be obtained by -
(a) respective additions of the individual parameters,
(b) respective multiplications of the individual parameters,
(c) matrix product of the individual parameters,
(d) none of the above.

Q.54 In electrical circuit analysis, state variables are -
(a) current for inductors and voltage for capacitors,
(b) current for capacitors and voltage for inductors,
(c) current or voltage for both inductors/capacitors,
(d) none of the above.

Q.55 In electrical circuit nalysis, a tree is a path covering all the nodes and having -
(a) no closed path, (b) only one closed path,
(c) maximum loops, (d) none of the above.

Q.56 In electric circuit analysis, if f(t) = f(-t), the functional symmetry is called -
(a) odd symmetry, (b) even symmetry,
(c) odd-even symmetry, (d) none of the above.

Q.57 In electric circuit analysis, a function is said to have odd symmetry if -
(a) f(t) = f(-t), (b) f(t) = -f(-t),
(c) f(t) = 1/f(t), (d) none of the above.

Q.58 The product of two functions having the same type (odd or even) symmetry, is a function having -

(a) odd symmetry, (b) even symmetry,
(c) odd or even symmetry, (d) none of the above.

Q.59 The product of two functions having different types symmetries (one odd and the other even) is a function having -

(a) odd symmetry, (b) even symmetry,
(c) odd or even symmetry, (d) none of the above.

Q.60 Fourier series of a function with even-symmetry will have only -

(a) a constant and a Cos function,
(b) a Sin function,
(c) a constant and a sin function,
(d) none of the above.

Q.61 Fourier series of a function having odd symmetry, will not have -

(a) a constant and a Cos function,
(b) a Sin function,
(c) a constant and a sin function,
(d) none of the above.

Q.62 In engineering mathematics, convolution, infact, means -

(a) integration of the product, (b) product of integrals,
(c) integral of division, (d) none of the above.

Q.63 A system function may be expressed as the ratio of (i) ratio of fourier transform of response function to fourier transform of forcing function, (ii) fourier transform of impuse response.

(a) both the above are true, (b) only (i) is true,
(c) only (ii) is true, (d) none of the above is true.

Q.64 The system function of a network can be found out to be -

(a) same as transfer function, (b) sinusoidal transfer function,
(c) any of the above two, (d) none of the above.

Q.65 The system function of a network is -

(a) fourier transform of unit-impulse response,
(b) laplace transform of unit step response,
(c) any of the above,
(d) none of the above.

Q.66 A 2 ohm resistor having 1 ampere current will dissipate power of -

(a) 2 juoules, (b) 2 joules/sec,
(c) 1/2 Joule, (d) none of the above.

Q.67 When 2 volt supply is put across a 1 farad capacitor, its energy storage will be -

(a) 2 joules, (b) 2 watts,
(c) 4 joules, (d) 1 joule.

Q.68 Energy storage in a 1 henry inductance supplied with a current of 2 amperes, will be -

(a) 2 watts, (b) 2 joules,
(c) 4 watts, (d) 4 joules.

Q.69 Number of loop equations for an electrical circuit with 10 elements and 7 nodes will be -

(a) 10, (b) 4,
(c) 3, (d) 7.

Q.70 If two inductors having self-inductance L1 and L2 and a mutual-inductance M, are connected in series, with opposite polarity, the total inductance of the combination will be -

(a) L1 + L2 - 2M, (b) L1 - L2 + 2M,
(c) L1 + L2 + 2M, (d) L1 + L2 - M.

Q.71 In parallel circuit resonance, the following parameter gets magnified -

(a) current, (b) voltage,
(c) power, (d) none of the above.

Q.72 The power factor of a RLC parallel circuit at resonance, will be -

(a) zero, (b) 0.5 lagging,
(c) 0.5 leading, (d) none of the above.

Q.73 A circuit, whose any of the parameter can be reversed and still the circuit works, is called -

(a) reversible, (b) bilateral,
(c) toggle, (d) none of the above.

Q.74 As compared to the original current drawn by a series RLC circuit, the current at resonance will be -

(a) lower, (b) higher,
(c) same.

Q.75 As compared to the original current drawn by a parallel RLC circuit, the current at resonance will be -

(a) lower, (b) higher,
(c) same.

Q.76 Examples of a current source are -

(a) a dc generator and a current transformer,

(b) an arc welder and a potential transformer,
(c) an arc welder and a current transformer,
(d) all of the above,

Q.77 Examples of a dependent voltage source, are -
(a) armature of a dc generator and an alternator,
(b) a dc generator and an alternator,
(c) a dc generator armature and an alternator armature,
(d) all the above,

Q.78 Duals of voltage, capacitance and loop are respectively -
(a) voltage, inductance and node,
(b) current, resistance and node,
(c) current, inductance and node,
(d) none of the above.

Q.79 In the superposition theorem, the following can be superposed -
(a) voltage sources only,
(b) voltage and current sources,
(c) voltage source and an inductance,
(d) none of the above.

Q.80 Conjugate of (A + jB) is -
(a) A -jB, (b) -A - jB,
(c) -A + jB, (d) none of the above.

Q.81 A current wave is represented by the equation i = 10 Sin 6.28 t. What is the supply frequency -
(a) 1 Hz, (b) 2 Hz,
(c) 6.28 Hz, (d) none of the above.

Q.82 Heating effects of an ac current with average value of 10 ampere and a dc current of 10 ampere will be -
(a) same,
(b) different,
(c) can't be determined with this data,
(d) none of the above.

Q.83 A very high capacitance, nearly behaves as -
(a) short circuit to ac and dc,
(b) short-circuit to ac and open-circuit to dc,
(c) short-circuit to dc and open-circuit to ac,
(d) none of the above.

Q.84 As compared to the r.m.s. value, the average value of a purely sinusoidal wave is -
(a) same, (b) lower,
(c) higher.

Q.85 For two series-connected resistors, the power dissipation is proportional to -

(a) square of their voltage drops,
(b) resistance values,
(c) inverse of resistance values,
(d) none of the above.

Q.86 Two resistors connected in series and fed from a voltage source, each dissipate a power of 10 watts, are reconnected in parallel, power dissipation in each will be -

(a) 5 watt, (b) 20 watt,
(c) 40 watt, (d) none of the above.

Q.87 Two resistors connected in series and fed from a current source, each dissipate a power of 10 watts, are reconnected in parallel, power dissipation of each will be -

(a) 2.5 watt, (b) 5 watt,
(c) 20 watt, (d) none of the above.

Q.88 A resistance wire of 180 ohms is trippled and ends short-circuited, its resistance will become -

(a) 60 ohm, (b) 20 ohm,
(c) 11.25 ohm, (d) none of the above.

Q.89 In circuits, transients are caused by -

(a) resistance only, (b) inductance only,
(c) reactance only, (d) none of the above.

Q.90 In ionized gases, the current flow depends upon -

(a) ions only, (b) electrons only,
(c) ions and electrons, (d) none of the above.

Q.91 Two resistors each of 40 ohm, 1 watt capacity are joined in parallel. Their effective rating will become -

(a) 40 ohm, 2 watt, (b) 20 ohm, 2 watt,
(c) 20 ohm, 4 watt, (d) none of the above.

Q.92 Four permanent bar magnets are forming a cross. Their respective flux-lines -

(a) will cross each other,
(b) will be combined together,
(c) will cross at some places and combine at some places,
(d) none of the above.

Q.93 Magnetic flux-lines can never be killed or crossed by other magnetic flux-lines.

(a) true, (b) false,
(c) partially true, (d) nonsense.

Q.94 A solenoid means -
(a) a solid magnet,
(b) a coil on a bar magnet,
(c) a coil to make a bar magnet,
(d) none of the above.

Q.95 When a voltage wave reverses, the current wave produced by it also reverses -
(a) always instantly, (b) never instantly,
(c) may be instantly, (d) none of the above.

Q.96 A higher electrical load means -
(a) higher resistance,
(b) lower resistance,
(c) same resistance but higher voltage,
(d) none of the above.

Q.97 Siemens is the unit of -
(a) flux, (b) conductance,
(c) parmeance, (d) none of the above.

Q.98 An electrical load has in series a resistance of 40 ohms and a reactance of 40 ohms, its power factor will be -
(a) 0.5, (b) 0.714,
(c) 0.367, (d) none of the above.

Q.99 An electric arc is -
(a) resistive always, (b) never resistive,
(c) sometimes resistive, (d) none of the above.

Q.100 As the frequency is increased, a capacitive reactance -
(a) increases, (b) decreases,
(c) remains same.

Q.101 With parallel connections of identical capacitors, -
(a) capacitance increases and reactance decreases,
(b) capacitance and reactance decrease,
(c) capacitance decreases and reactance increases,
(d) none of the above.

Q.100 As the frequency is increased, a capacitive reactance -
(a) increases, (b) decreases,
(c) remains same.

Q.101 With parallel connections of identical capacitors, -
(a) capacitance increases and reactance decreases,
(b) capacitance and reactance decrease,

(c) capacitance decreases and reactance increases,
(d) none of the above.

Q.102 A bulb connected in series with a capacitor and connected to a dc source, will -
(a) flicker at resonance frequency,
(b) not glow at all,
(c) will glow less,
(d) none of the above.

Q.103 The electrical energy supplied to our homes in India is measured and billed on the basis of units consumed. The unit means -
(a) kWh, (b) kW,
(c) Newton/s, (d) none of the above.

Q.104 The The rating of batteries is given in -
(a) kVA, (b) kWh,
(c) Ah, (d) none of the above.

Q.105 The battery rating is usually obtained by -
(a) no. of cells in it,
(b) its cabinet size,
(c) discharging it thru an instrument,
(d) none of the above.

Q.106 A series RLC circuit is fed with an ac supply at resonant frequency, the sum of the voltages measured separately across each element, will be -
(a) equal to the supply voltage,
(b) less than the supply voltage,
(c) more than the supply voltage.

Q.107 Three equal resistors connected in parallel are fed with a battery, if one of them is open cicuited,
(a) current in other resistors will jump to 150 percent,
(b) voltage across other resistors will increse to 150 percent,
(c) current and voltage across other resistors will not change,
(d) none of the above.

Q.108 In a gas, current can flow through the movement of -
(a) ions only, (b) electrons only,
(c) both ions and electrons, (d) none of the above.

Q.109 Earth magnetic axis is -
(a) parallel to the equator, (b) perpendicular to equator,
(c) independent of equator, (d) none of the above.

Q.110 Earth's magnetic effect is due to -
(a) hidden magnetic material inside it,
(b) its spinning,
(c) its orbitting around sun,
(d) none of the above.

Q.111 A resonant circuit of parallel capacitor and inductor combination, connected across an ac supply of resonant frequency, will draw power from the source -
(a) in huge quantity - ideally infinite,
(b) in very low quantity - ideally zero,
(c) depending upon the impedance of any one component,
(d) none of the above.

Q.112 A series RLC circuit has resistance R, and reactance X. Its total impedance is Z, its conductance will be -
(a) R/Z^2, (b) X/Z^2,
(c) $1/R$, (d) none of the above.

Q.113 A series RLC circuit has resistance R, and reactance X. Its total impedance is Z, its succeptance will be -
(a) R/Z^2, (b) X/Z^2,
(c) $1/R$, (d) none of the above.

Q.114 A series RLC circuit has resistance R, and reactance X. Its total impedance is Z, its admittance will be -
(a) R/Z^2, (b) X/Z^2,
(c) $1/R$, (d) none of the above..

Q.115 When the applied voltage to a series RLC circuit is 1 volt, its conductance will be equal to the -
(a) current in phase with the voltage,
(b) current lagging the voltage by 90 degrees,
(c) total current,
(d) none of the above.

Q.116 When the applied voltage to a series RLC circuit is 1 volt, its susceptance will be equal to the -
(a) current in phase with the voltage,
(b) current lagging the voltage by 90 degrees,
(c) total current,
(d) none of the above.

Q.117 When the applied voltage to a series RLC circuit is 1 volt, its admittance will be equal to the -
(a) current in phase with the voltage,
(b) current lagging the voltage by 90 degrees.

(c) total current,
(d) none of the above.

Q.118 The total heat produced by a heater coil of 5 ohm resistance ,as compared to that produced by a thick copper rod of 5 ohm resistance will be -

(a) the same, (b) will be very much higher,
(c) will be very much lower, (d) none of the above.

Q.119 In a domestic ceiling fan, the armature of the induction motor is on the -

(a) outer side, (b) inner side,
(c) on top, (d) at the bottom.

Q.120 If two domestic heaters of 1000 watt each are connected in series, and fed with domestic supply, the total heat produced by them will be -

(a) 2000 w, (b) 500 w,
(c) 250 w, (d) none of the above.

Q.121 The main purpose of providing lead sheath over the cables is -

(a) mechanical strength, (b) earth connection,
(c) hygroscopic protection, (d) none of the above.

Q.122 A micrometer gauge is used to read -

(a) microamperes, (b) pressure,
(c) periphery, (d) distance,

Q.123 For a particle of mass m, moving with velocity v, at a radius r, from its center of rotation, its momentum will be -

(a) mv, (b) mvr,
(c) mv/r, (d) none of the above.

Q.124 For a particle of mass m, moving with velocity v, at a radius r, from its center of rotation, its moment of momentum will be -

(a) mv, (b) mvr,
(c) mv/r, (d) none of the above.

Q.125 For a particle of mass m, moving with velocity v, at a radius r, from its center of rotation, its angular momentum will be -

(a) mv, (b) mvr,
(c) mv/r, (d) none of the above.

Q.126 The force of attraction between two opposite magnetic poles is proportional to -

(a) the distance between them,
(b) the inverse of distance between them,
(c) the square of the inverse of the distance between them,
(d) none of the above.

Q.127 If in a conductor rate of change of current flow is 1 A/s and the induced emf in the coductor is 2 volts, the self-induction ofthe conductor will be -

(a) 2 farad, (b) 2 henry,
(c) 1/2 henry, (d) none of the above.

Q.128 If in a conductor rate of change of current flow is 1 Amp per half second and the induced emf in the coductor is 1 volts, the self-induction ofthe conductor will be -

(a) 2 farad, (b) 2 henry,
(c) 1/2 henry, (d) none of the above.

Q.129 A permanent magnet type of voltmeter is connected across a 100 volt ac supply. The reading of the meter will be -

(a) 100 V, (b) 144 V,
(c) 100/1.44 V, (d) none of the above..

Q.130 With the flow of an ac in a conductor, a magnetic flux is produced and an emf is induced. The phasor lagging behind the current phasor by 90 degrees, will be that of -

(a) magnetic flux,
(b) induced emf,
(c) both magnetic flux and induced emf,
(d) none of the above.

Q.131 Thc ac current is measured in rms value for its equivalent power consumption compatibility, the voltage is measured in rms value, basically for its compatibility with -

(a) equivalent power consumption,
(b) current measurement,
(c) equivalent energy consumption,
(d) none of the above.

Q.132 A condenser, fed with a dc voltage V, holds a charge Q, its capacity will be -

(a) Q/V farad, (b) Q/V microfarad,
(c) QV farad, (d) none of the above.

Q.133 When an idcal condenser is connected across an ac supply, the current flow will be -

(a) through the dielectric,
(b) externally from plate to plate,
(c) through both (a) and (b) above,
(d) none of the above.

Q.134 In a sinusoidal wave, instantaneous value of the following varies is sinusoidal -

(a) voltage, (b) voltage and current,
(c) power, (d) voltage, current and power.

Q.135 A sinusoidal wave can also be called a -

(a) cosine wave, (b) tangent wave,
(c) cotangent wave, (d) none of the above.

Q.136 In an ac if average values of current and voltage are zero, the average value of power will be -

(a) zero, (b) positive,
(c) negative, (d) none of the above.

Q.137 For an ac current of frequency 50 Hz, the frequency of power variation wave will be -

(a) 50, (b) 100,
(c) zero, (d) none of the above.

Q.138 Electrostatic shielding of an equipment can be done by -

(a) enclosing it in a dielectric box,
(b) enclosing it in a metal box,
(c) connecting it to earth,
(d) none of the above.

Q.139 The charge on an conducting sphere stays at its outer surface only due to -

(a) attraction by the surroundings,
(b) repulsion from the center,
(c) mutual repulsion of charge elements,
(d) none of the above.

Q.140 If a charged capacitor is shunted by an inductor, the rsulting current may be -

(a) dc constant, (b) oscillatory constant,
(c) oscillatory decreasing, (d) none of the above.

Q.141 Current capacity of a large-size conductor can be taken as equal to (k times its area of cross-section), where k is a constant with value -

(a) one, (b) less than one,
(c) more than one,

Q.142 As compared to overhead transmission lines, cables have -

(a) higher series impedance, (b) higher series conductance,
(c) higher shunt conductance, (d) none of the above.

Q.143 Resistance of a conductor, with temperature coeft.- (α), and resistance at 20 degree C as R, at any temperature t is given by -

(a) R [α.(t-20)], (b) R [1+α.(t-20)],
(c) R [1 + αt], (d) none of the above.

Q.144 The charge density on the surface of a conductor is maximum at -

(a) the flat surface, (b) the spherical surfaces,
(c) the pointed tips, (d) none of the above.

Q.145 By adding tin to copper its conductivity is -

(a) increased, (b) decreased,
(c) not affected,

Q.146 As compared to the conductivity of mild steel, conductivity of sodium is -

(a) higher, (b) lower,
(c) almost same, (d) incomparable.

Q.147 The International Annealed Copper Standard (IACS) is established to get relative resistivity of conductors as compared to the resistivity of -

(a) a superconductor, (b) copper at 0 degree celsius,
(d) copper at 20 degree C, (d) none of the above.

Q.148 In overhead lines of copper conductors, tension and the compaction may reduce the conductance by nearly -

(a) 0.5 - 1.0 percent, (b) 2.5 -3 percent,
(c) 4 - 5 percent, (d) none of the above.

Q.149 As conductor material, copper is being replaced by aluminium due to -

(a) supply problem, (b) economy in use,
(c) reasons of conductivity, (d) none of the above.

Q.150 Alumiuium conductivity is around -

(a) 60 % IACS, (b) 1/3 of copper,
(c) 50 % of superconductor, (d) none of the above.

Q.151 Density of aluminium is around -

(a) 1/3 of copper density, (b) 1/4 of steel density,
(c) twice the water density, (d) none of the above.

ABSTRACT

1.1. SI SYSTEM OF UNITS

(a) Basic quantities

Quantity	*Unit*	*Symbol*
Length	Meter	m
Mass	Kilogram	kg
Time	Second	s
Electric current	Ampere	A
Thermodynamic temperature	Kelvin	K
Amount of substance	Mole	mol
Luminous intensity	Candela	cd

(b) Supplementary quantities

1. Plain angle (radians)

One radian is the angle between two radii of a circle, which cut off on the circumference an arc equal in length to the radius.

2. Solid angle (steradian)

One steradian is the solid angle at the vertex of a cone with its vertex at the centre of a sphere and cutting off an area on the surface of the sphere, equal to that of a square with sides equal to the radius of the sphere.

(c) Experimentally verified units

Name	*Symbol*	*Approximation*
Electron volt	eV	1 eV = 1.60219×10^{-19} J
Unified atomic mass unit	u	lu = 1.66053×10^{-27} kg
Astronomical unit	AU	1 AU = 1.496×10^{11} m
Parsec	pc	1 pc = 206,264.8 AU = $30,857 \times 10^{12}$ m

(d) Derived electrical units

Quantity	*SI units*			
	Name	*Symbol*	*Expression in base units*	*Usual expression*
Frequency	hertz	Hz	S^{-1}	1/s
Force	newton	N	$m.kg.s^2$	N
Pressure, stress	pascal	Pa	$kg.m^{-1}.s^{-2}$	N/m^2

(Contd.)

Quantity	*SI units*			
	Name	*Symbol*	*Expression in base units*	*Usual expression*
Energy, work, heat	joule	J	kg.m^2.s^{-2}	N.m
Power, radiation	watt	W	kg.m^2.s^{-3}	J/s
Electricity, charge	coulomb	C	A.s	A.s
Potential difference, EMF, voltage	volt	V	kg.m^2.s^{-3}A^{-1}	W/A
Electric capacitance	farad	F	A^2.s^4.m^{-2}.kg^{-1}	C/V
Electric resistance	ohm	W	kg.m^2.s^{-3}.A^{-2}	V/A
Electric conductance	siemens	S	A^2.s^{-3}.kg^{-1}.m^{-2}	A/V
Magnetic flux	Weber	wb	kg.m^2.s^{-2}.A^{-1}	V.s
Magnetic flux density	testa	T	kg.s^{-2}.A^{-1}	Wb/m^2
Temperature (celsius)	degree celcius	°C	-	K
Inductance	hurry	H	kg.m^2.s^{-2}.A^{-2}	Wb/A
Luminous flux	lumen	lm	cd.sr	-
Illuminance	lux	lx	cd.sr.m^{-2}	lm/m^2
Radio nuclide activity	beequerel	Bq	S^{-1}	1/s
Absorbed dose	gray	Gy	m^2.s^{-2}	J/kg
Dose equivalent	sievert	Sv	m^2.s^{-2}	J/kg

1.2. BASIC ELECTRICITY

All materials are built of atoms - a negatively charged electron cloud around a positively charged nucleus. The displacement of charge is called flow of electricity. Charge of a single electron is - 1.602 × 10^{-19} coulombs.

Voltage

When the charge is polarized between two points, a voltage or potential difference is said to exist between the two points. It is the basic cause of flow of charge. It is measured in volts (V).

Current

Flow of charge is electric current. If a material capable of being a path of charge is placed between two points with a voltage, a current is said to be flowing between the two points for positive potential to negative potential. It is measured in amperes (A).

Resistance

Hinderance to the flow of electrons in a given material is said to be its resistance. It is measured in ohms (Ω). Reciprocal of resistance is known as conductance. Its unit is Siemens.

Ohm's law

The voltage V between two points of a current carrying electrical conductor is directly proportional to the current flowing in it, i.e. V = RI where R is the proportionality constant, known as the resistance of the material in which the current is flowing.

Resistivity

It is the characteristic property of the material of a conductor determining the resistance of particular specimen of the material by the relationship :

$$R = \rho \frac{l}{A}$$

where R = resistance of the specimen
l = length of specimen in the direction of flow of current
A = area of the specimen perpendicular to the direction of current flow
l = resistivity of the material.

Thus resistivity of a material can be said to be the resistance of a specimen of unit length and unit cross-section.

Resistivity and hence the resistance of a conductor is dependent upon its temperature and is related by the equation :

$$\frac{R_2}{R_1} = \frac{T - t_1}{T - t_2}$$

where R_2 = resistance at temperature t_2°C
R_1 = resistance at temperature t_1°C
T = absolute temperature at which the resistivity of the material, theoritically, is to be zero.

Another relationship showing the temperature dependence of resistance is :

$$R_2 = R_1[1 + \alpha (t_2 - t_1)]$$

where α is called the temperature coefficient of resistance for the material at the operating temperature.

Metals have a positive temperature coefficient of resistance while the semiconductors and insulators have a negative temperature coefficient of resistance.

Resistances in combinations

(i) For resistances R_1, R_2 . . . connected in series, equivalent resistance R is given by :
$R = R_1 + R_2 + \ldots$
(ii) For resistances R_1, R_2 . . . connected in parallel, the equivalent resistance R is given by :

$$\frac{1}{R} = \frac{1}{R_1} + \frac{1}{R_2} + \cdots$$

1.3. ELECTROSTATICS

When an electric voltage is connected to two points between which there is no flow of current, an electrostatic field is said to exist between the two points. A quantity known as electric flux is said to be flowing between the two points.

Electric field intensity (E)

Electric field intensity or the potential gradient between two points is the ratio of electrostatic potential difference and the distance between the two points.

Electric flux density (D)

Electric flux density is the ratio of lines of electric flux to the area of cross section.

Permitivity (ε)

It is the characteristic of a dielectic material, which permits the flow of flux through it. It is given by :

$$\varepsilon = \frac{D}{E} \text{ F/m}$$

and $\varepsilon = \varepsilon_o \varepsilon_r$

where $\varepsilon_o = 8.854 \times 10^{-12}$ F/m, permitivity of free space

ε_r = Relative permitivity of the material

Capacitor

Two conducting elements separated by a dielectric form a device known as capacitor. The capacity of the capacitor to hold electric charge per unit viltage is known as capacitance. Thus :

$$C = \frac{Q}{V} \text{ Farad (F)}$$

where Q = charge in coulombs

V = voltage in volts

The instantaneous charging current of a capacitor is given by :

$$i = C\frac{dV_c}{dt}$$

where V_c is the voltage across the capacitor plates.

The energy stored by a capacitor is given by :

$$w = \tfrac{1}{2}CV^2$$

The value of capacitance of a parallel plate capacitor is :

$$C = \varepsilon \frac{A}{d}$$

where ε = permitivity of the dielectric
A = area of the conducting plates
d = distance between the plates

Capacitor combinations

(i) When capacitors of capacitances C_1, C_2, . . . are connected in series, the equivalent capacitance (C) is given by :

$$\frac{1}{C} = \frac{1}{C_1} + \frac{1}{C_2} + \ldots$$

(ii) When capacitors of capacitances C_1, C_2, . . . are connected in parallel, the equivalent capacitance (C) is given by :

$$C = C_1 + C_2 + \ldots$$

1.4. ALTERNATING CURRENT (AC)

When voltage and hence the current in an electric system vary sinusoidally with time, the system is said to be an AC system. The voltage or current wave form is given by the equation

$v(t) = V_m \sin \omega t$ and $i(t) = I_m \sin \omega t$
or $v(t) = V_m \cos \omega t$ and $i(t) = I_m \cos \omega t$

where $v(t)$ is the instantaneous value of voltage
V_m is the peak value of amplitude
ω is the angular frequency
$= 2\pi f$, where f is the frequency in Hz.

Angle

Angle is usually measured in degrees. A complete circle has 360 degrees. In engineering radian is a more popular and convenient unit, where

$$\text{No. of radians in a circle} = \frac{\text{circumference}}{\text{radius}} = 2\pi$$

or 2π radians = 360°
1 radian = 57.3°

Average value

Average value of a sinusoidal wave form per half cycle = 0.637 x Maximam value

Effective value (RMS value)

Effective or RMS value of a sinusoidal wave
= Maximum value/√2
= 0.707 x Maximum value.

Form factor

For a sinusoidal wave, form factor $= \dfrac{\text{RMS value}}{\text{Mean value}} = 1.11$

Fundamental and harmonic components

Every periodic wave can be resolved into a number of sinusoidal waves. The prime of these waves with minimum of frequency is called the fundamental component and the sinusoidals of higher frequencies are called the harmonics. The frequencies of harmonics are integral multiples of fundamental frequency.

Phasor and phasor angle

A sinusoidally varying quantity can conveniently be represented by a vector of length equal to the rms value and at angle, with respect to any reference axis, equal to the angle on time scale from 0 to 2π for each cycle. Thus $V<\theta$ is the phasor representation of a voltage wave giving the instataneous value at the position of angle θ.

Various interrelated waves may be plotted on common reference axes, showing the time interrelations of the quantities, such diagrams are called phasor diagrams.

Three-phase system

For economic reasons, AC power systems are of 3 phases equilaterally spaced in time. The three phases of an electrical system may be connected in Wye or Delta connections. The star connection provides a fourth terminal also known as the neutral. The three main terminals are known as the line terminals.

For a Wye connection :

Line-to-neutral voltage $= \dfrac{1}{\sqrt{3}} \times$ line-to-line voltage

Line current = Phase current.

For a Delta connection :

Line voltage = Phase voltage

Line current = √3 phase current

Magnetic circuits

A current carrying conductor produces a magnetic field around it. A magnetic field is represented by lines of magnetic flux, emanating from north pole to south pole and never intersecting each other.

If a coil of N turns carries a current I, the magnetomotive force produced by the coil is NI ampere turns producing a magnetic flux in the medium, the quantity of which depends upon the reluctance of the medium. The reluctance of material is given by :

$$R = \frac{l}{\mu A}$$

where l is the length of path

μ is the magnetic permeability of the material

$= \mu_o \mu_r$ = magnetic permeability of space x relative permeability

$= 4\pi \times 10^{-7}$ x relative permeability

The magnetic flux density (B) is given by :

$$B = \frac{\Phi}{A} = \mu H = \mu \frac{NI}{l} \text{ wb/m}^2$$

where Φ is the total flux

A is the area of cross section

H is the magnetic field intensity in A/m

Magnetic flux density and field intensity are related by B-H loop, the area of which represents the power loss per cycle. The relationship between the magnetizing force and the magnetic flux density deviates from being linear for the two main reasons.

(i) Magnetic saturation

Materials do magnetise because of the permanent magnetic dipoles in the materials are aligned with each other producing magnetic effect. As the magnetization increases, lesser dipoles remain in the material for alignment, hence more magnetizing effort becomes necessary to get the same magnetization as was possible at lower levels of magnetization. A situation comes when all the dipoles of the material are aligned and the magnetization of the material becomes constant, no matter what the magnetizing effort is present. This situation is called magnetic saturation.

(ii) Magnetic leakage

When an electric circuit is used to magnetize a magnetic circuit, not all flux is produced in the magnetic material but some flux is leaked to unwanted circuits or to the surrounding air. This leakage of flux is more if the electric and the magnetic circuits are loosely coupled. In electrical circuits and machines this leakage flux causes leakage reactance.

The ratio of the total flux to the useful flux is called the magnetic leakage coefficient.

Rules of electromagnetism

The important rules governing the electromagnetism are :

(i) Cork-screw rule

This rule determines the direction of flux produced by a current. If we imagine a cork-screw to point in the direction of current flow in a conductor and we intend to move the screw forward in the direction of current flow, the direction of rotation of the cork screw head will be the direction of the flux flow around the conductor.

(ii) Magnetic pole forming rule

This rule determines the magnetic pole formed by the flow of current in a coil. Imagine the capital letters N and S and put arrow-tips on the free ends of the alphabets, resulting in the look of rotational torques in two directions from the two letters. N provides an anticlockwise rotation while S shows a pattern of clockwise rotation. These patterns are the direction of currents in the coil to form the respective magnetic poles, i.e. a current in the anticlockwise direction will form the north pole on the coil axis upper end, while a clockwise current in the coil will form a south pole on top of the coil.

(iii) Faraday's law of electromagnetic induction

'Whenever a conductor cuts magnetic lines of flux, a voltage is induced in the conductor'. This is Faraday's first law.

The second law of Faraday states that the magnitude of the induced emf is proportional to the rate of change of the magnetic flux linking with the conductor or circuit. Mathematically, if a conductor of length l moves in a uniform flux density B, perpendicular to it, with a velocity v, the induced voltage in the conductor will be Blv volts.

If the flux distribution is sinusoidal and the conductor moves by an angle θ from the line of action of the magnetic field, the induced voltage will be $Blv \sin\theta$.

(iv) Fleming's right hand rule

This rule determines the direction of the induced emf, by electromagnetic induction. Arrange your thumb, first finger and the second finger of your right hand to point towards the three mutually perpendicular directions. If the thumb points towards the direction of motion and first finger towards the direction of magnetic field, the second finger will point towards the direction of the induced emf.

(v) Law of electromagnetic interaction

If a current carrying conductor is put in a magnetic field perpendicular to it, the conductor will experience a force on it.

Most electric motors work on this principle.

(vi) Fleming's left hand rule

This rule gives the direction of force produced due to electromagnetic interaction.

Arrange thumb, the first and the second fingers of your left-hand in the three mutually perpendicular directions. If the first finger points towards the field and the second finger towards the direction of flow of current, the thumb will indicate the direction of the force on the conductor.

(vii) Lenz's law

In general terms, this rule states that in electromagnetic induction, the effect produced by a cause, will tend to oppose the cause. In specific terms for the induced emf, the direction of the induced emf will be such that any current resulting from it will produce a magnetic flux will oppose the change of flux linkage or motion, inducing the emf.

Self-inductance

A change of magnitude of current in a conductor produces a proportional change of flux linking the conductor, inducing an emf in the conductor. The induced emf is proportional to the rate of change of current. Mathematically :

$$e = -L \frac{di}{dt}$$

where L is the constant of proportionality depending upon the circuit and known as the self-inductance of the circuit.

$$\text{since } e = -N \frac{d\Phi}{dt}$$

$$L = N \frac{d\Phi}{di}$$

$$= N \frac{\Phi}{I}$$

where N is the number of turns in the coil.

Also for a solenoid of length l and area of cross-section a, having N turns:

$$L = \mu \frac{aN^2}{l}$$

where μ is the permeability of the medium inside the solenoid. Self-inductance can never have a negative value.

Mutual inductance

When two coils are mutually coupled, rate of change of current in one coil induces an emf in the other coil.

Thus $e_1 = -M_{12} \dfrac{di_2}{dt}$

where M_{12} is proportionately constant known as the mutual inductance of coil one with coil two, similarly :

$$e_2 = -M_{21}\frac{di_1}{dt}$$

since $e_1 = -N_1 \dfrac{d\Phi_{21}}{dt}$

where Φ_{21} is the flux linking both the coils.

hence $M_{12} = N_1 \dfrac{d\Phi_{21}}{di_2}$

and $M_{21} = N_2 \dfrac{d\Phi_{12}}{di_1}$.

$$\text{If } N_1 = N_2 = N,\ M = M_{12} = M_{21} = N\frac{\Phi_{12}}{i_1} = N\frac{\Phi_{21}}{i_2}$$

$$\text{Coupling coefficient } k = \frac{\text{flux linking } N_1 \text{ and } N_2}{\text{total flux}}$$

$$= \frac{M}{\sqrt{L_1.L_2}}$$

RLC circuits

(i) Inductive circuit

For a current $i = I_m \cos \omega t$

$$\text{voltage induced } v = -L\frac{di}{dt}$$

$$= \omega L.I_m. \sin \omega t$$

$$= \omega L\ I_m \cos (\omega t + 90^o)$$

$$|v| = \omega L\ i$$

hence $\dfrac{v}{i} = \omega L = 2\pi f L = X_L$

X_L is known as the inductive reactance of the circuit. The above relations between v and i also shows that the voltage is an inductive circuit leads the current by 90°.

The power in an inductive circuit

$P_L = v.i.$

$= V_m \sin \omega t \, I_m \cos \omega t$

$= V_m I_m \sin 2\omega t$ watts

Hence the instantaneous power is positive for quarter cycle and negative for another quarter cycle of the current wave. Thus the inductance continuously draws and returns power to the source, such that the average power for the full cycle of the current is always zero.

(ii) Capacitive circuit

For an applied voltage $v = V_m \cos \omega t$

The instantaneous current is given by :

$$i = C \frac{dv}{dt}$$

$$= \omega C \, v_m \sin \omega t$$

$$= \omega C \, v_m \sin \omega t$$

$$= \omega C \, v_m \cos (\omega t + 90^o)$$

$$= \omega C v$$

$$\text{Hence } \frac{v}{i} = \frac{1}{\omega C} = \frac{1}{2\pi f C} = X_c$$

where X_c is the capacitive reactance. Also from above, the voltage wave always lags the current wave in a capacitance circuit.

The instantaneous power in a capacitive circuit is :

$p_c = V_m I_m \cos \omega t \sin \omega t$

$= V_m I_m \sin 2\omega t$ watt

Hence the capacitor draws power for the quarter cycle of voltage wave and returns it back to the source in the next quarter cycle. The average power for the whole cycle is zero.

(iii) Series RLC circuit

Consider the three elements, a resistor R, an inductor L and a capacitor C connected in series across a sinusoidal voltage. Let i is the common current ad v_R, v_L and v_i are the voltages across the three elements.

$$v = v_R + j\,(v_L - v_C)$$

$$= iR + j(2\pi f L.i + \frac{1}{2\pi f C}.i)$$

$$\frac{v}{i} = R + j(2\pi f L + \frac{1}{2\pi f C})$$

$$= R + j(X_L - X_C) = Z$$

where Z is the impedance of the circuit with magnitude :

$$|Z| = \sqrt{R^2 + (X_L - X_C)^2}$$

The current will lag the voltage by an angle

$$\Phi = \tan^{-1}\frac{(X_L - X_C)}{R}$$

called the power factor angle of the circuit.

(iv) Parallel RLC circuit

In the analysis of series RLC circuit, if we replace reactances by susceptances and the resistance by conductance, the impedance is replaced by admittance and we get the result.

$$\text{Admittance } |Y| = \sqrt{R^2 + (B_L - B_C)^2}$$

$$\text{and the power factor angle } \Phi = \tan^{-1}\frac{(B_L - B_C)}{G}$$

$$\text{where } G = \frac{1}{R},\ B_L = \frac{1}{\omega_L} \text{ and } B_C = wC$$

(v) Resonance in RLC circuits

From the above analysis, we have noticed that :

(i) the inductive reactance increases with increase in frequency.

(ii) the capacitive reactance decreases with increase in frequency.

(iii) and the two work opposite to each other because of the mutually opposite phase angles.

Therefore if an RLC circuit is fed with a supply of variable frequency, a value of frequency will be such that :

$X_L = X_C$ or $B_L = B_C$

In such a situation the circuit behaves as a purely resistive circuit. This type of situation is called the resonance of the circuit. It occurs in both, the parallel and the series circuits.

1.5. CIRCUIT THEORY

Kirchoff's voltage law (KVL)

In a planar circuit, the algebraic sum of potential differences across all the elements of a close loop is zero, i.e. :

$\Sigma v_n = 0$

Kirchoff's current law (KCL)

The algebraic sum of the currents entering or leaving a node is zero, i.e. :

$\Sigma\, i_n = 0$

Superposition theorem

The total component voltage or current in a multi emf source circuit is the algebraic sum of the individual voltages or currents resulting from each voltage source applied one at a time with all other voltage sources removed and replaced by their respective internal resistances.

Thevnin's theorem

Any linear bilateral two terminal network, consisting of fixed resistances (or impedences) and sources of emf, can be replaced by a single source of emf V_{th} in series with a single resistance (or impedance). Voltage of the new source being the voltage at the terminals, when the external load to it is removed. The value of the series resistance (or impedance) is equal to the equivalent resistance (or impedance), measured or calculated at the terminal pair with all independent sources deenergized and replaced by their internal resistances (or impedances).

Norton's theorem

It is a dual of Thevmin's theorem. It states that at a terminal pair, any linear network can be replaced by a current source in parallel with a resistance (or impedance). The current of the current source being equal to the current that flows through the terminals when short-circuited and the parallel resistance (or impedance) being equal to the equivalent resistance (or impedance) at the terminal pair with all the independent sources deenergized and replaced by their internal resistances.

Wye-delta conversions

If R_1, R_2, R_3 are the three leg resistances (or impedances) of a Wye network with respective terminals as A, B and C. Also for a Delta network, let the three leg resistances (or impedances) between terminals A-B, B-C and C-A be R_{ab}, R_{bc} and R_{ca}. The two circuits will be equivalent if :

$$R_1 = \frac{R_{cb}\,R_{ca}}{R_{ab} + R_{bc} + R_{ca}}$$

$$R_2 = \frac{R_{ab}\,R_{bc}}{R_{ab} + R_{bc} + R_{ca}}$$

$$R_3 = \frac{R_{bc}\,R_{ca}}{R_{ab} + R_{bc} + R_{ca}}$$

$$\text{or } R_{ab} = \frac{R_1R_2 + R_2R_3 + R_3R_1}{R_3}$$

$$R_{bc} = \frac{R_1R_2 + R_2R_3 + R_3R_1}{R_1}$$

$$R_{ca} = \frac{R_1R_2 + R_2R_3 + R_3R_1}{R_2}$$

Circuit simplication methods

(i) Branch current method

This method involves the following steps :

1. Assign branch current variables.
2. Show polarity signs across circuit elements.
3. Choose arbitrary tracing loops.
4. Write as many independent Kirchoff's voltage law equations as there are current variables.
5. Solve the simultaneous equations to obtain the branch currents.

(ii) Loop or mesh current method

This method of circuit analysis involves the following steps :

1. Indicate a loop current variable for each loop.
2. Indicate polarity signs of voltage drops across the circuit elements for the assumed loop currents.
3. Write Kirchoff's voltage law equations for each loop using the algebraic sum of loop currents through each branch element to determine the voltage drop.
4. The number of equations should be equal to the number of variables.
5. Solve the simultaneous equations to get the loop currents.
6. Calculate the branch currents.

(iii) Node voltage method

Take the following steps :

1. Convert all voltage sources to current sources.
2. Pick a reference node arbitrarily and assign all other nodes voltage variables V_1, V_2 etc.
3. Write expressions for all branch currents based upon the node voltages.
4. Apply Kirchoff's circuit law for each node and obtain the equations.
5. Solve the equations to get the node voltages.

2

Materials and Properties

Q.1 The mass of and charge of an electron are respectively -
(a) 9.107 10^{-31} kg and 1.602 10^{-9} coulomb,
(b) 9.107 10^{-31} kg and 1.602 10^{-9} coulomb,
(c) 1.602 10^{-31} kg and 9.107 10^{-9} coulomb,
(d) none of the above.

Q.2 Conductivity of a material depends upon -
(a) number of electrons, (b) number of protons,
(c) number of neutrons, (d) none of the above..

Q.3 The number of electrons in orbit (n) of an atom, are -
(a) $2n^2$, (b) less than $2n^2$,
(c) more than $2n^2$, (d) none of the above.

Q.4 Chemical properties of a material depend upon its -
(a) atomic number,
(b) outermost electron configuration,
(c) innermost electron configuration,
(d) none of the above.

Q.5 The hardest of solids are those -
(a) which have an ionic structure,
(b) in which atoms share valence electrons permanently,
(c) in which valence electrons are share but are mobile,
(d) none of the above.

Q.6 NaCl and KF are examples of -
(a) ionic crystals, (b) valence crystals,
(c) metallic crystals, (d) none of the above.

Q.7 Diamond, Silicon and Germanium are the examples of -
(a) ionic crystals, (b) valence crystals,
(c) metallic crystals, (d) none of the above.

Q.8 Van der Waals crystals are those in which -
(a) the atoms share valence electrons,
(b) the valence electrons are continuously in a flux of mobility,
(c) the outer electron orbit is very stable,
(d) none of the above.

Q.9 Helium, Argon, Neon etc. are the examples of -
(a) ionic crystals, (b) valence crystals,
(c) metallic crystals, (d) Van der Waals crystals.

Q.10 Van der Waals crystals have no mutual cohesion and so are usually -
(a) solid, (b) liquid,
(c) inert gaseous..

Q.11 The value of dielectric constant in MKS units, for free-space is -
(a) $4\pi \times 10^{-7}$, (b) 8.854×10^{-12},
(c) $1.602\ 10^{-19}$, (d) none of the above.

Q.12 The total electric flux emanating from a closed surface is equal to the total charge enclosed by the surface. This is -
(a) Faraday's law, (b) Kirchoff's law,
(c) Gauss law, (d) none of the above.

Q.13 Charge (Q) produces electric field density (D) over an area (A), which causes electric field strength (E). For a material with relative dielectric constant ε_r, these are related as follows -
(a) $Q = \varepsilon_r.D.A = \varepsilon_o.E.A$, (b) $Q = D.A = \varepsilon_o.\varepsilon_r.E.A$,
(c) $E = D.A = Q/\varepsilon_r$, (d) none of the above.

Q.14 Relative dielectric constant of vacuum is -
(a) zero, (b) one,
(c) infinity, (d) none of the above.

Q.15 Capacitance of two parallel plates each of area A, separated by a distance d, and the gap filled by a dielectric material of constant ε_r, is given by -
(a) $\varepsilon_r.A.d$, (b) $\varepsilon_r.A/d^2$,
(c) $\varepsilon_o.\varepsilon_r.A/d$, (d) none of the above.

Q.16 Dielectric constant for insulating materials is usually -
(a) lies between zero and one,
(b) more than one,
(c) can be negativem,
(d) none of the above.

Q.17 Dielectric constant usually decreases as the temperature rises, with its lowest value at -

(a) 927 degree celcius, (b) absolute zero temperature,
(c) Curie temperature, (d) none of the above.

Q.18 In presence of an electric field, dielectric materials oppose the flow of current through them by raising -

(a) dipole moments, (b) counter currents,
(c) counter emf, (d) none of the above.

Q.19 During the processof raising dipole moments against applied electric fields, the dielectric materals undergo -

(a) dimensional change, (b) temperature change,
(c) quantum change, (d) none of the above.

Q.20 Dielectric polarization causes mechanical distortion in all materials. Mechanical distortion causes polarization -

(a) always, (b) never,
(c) in some materials..

Q.21 Most of the materials possess the property of electrostriction, which means dielectric polarization causes mechanical distortion and mechanical distortion causes -

(a) polarization, (b) no polarization..

Q.22 If a dielectric polarization has caused mechanical distortion in one direction, in the opposite direction, the effect -

(a) will always be same, (b) may be same,
(c) will not be the same.

Q.23 The material in which a dielectric polarization, in the two opposite directions, causes mechanical distortions of opposite natures, the materials are known as -

(a) insulators, (b) electrostrictive,
(c) piezoelectric, (d) none of the above.

Q.24 Dielectric materials having a center of charge-symmetry in their lattice structure, exibit the property of -

(a) electrostriction, (b) magnetostriction,
(c) piezoelectricity, (d) none of the above.

Q.25 Piezoelectric materials lattice structures have -

(a) a center of charge-symmetry,
(b) an axis of charge symmetry,
(c) a diagonal charge symmetry,
(d) no charge symmetry..

Q.26 In piezoelectric materials, the process of mechanical stress producing diectric polarization is reversible, in -
(a) all directions,
(b) at a specific axis,
(c) at mutually perpendicular axes,
(d) none of the above.

Q.27 In piezoelectric materials, if the application of mechanical stress at a specific axis, is reversed, the resultant polarization -
(a) is not altered, (b) is also reversed,
(c) become zero, (d) none of the above.

Q.28 Through piezoelcctric materials, conversion of mechanical energy to electric energy is -
(a) non-reversible, (b) reversible,
(c) impossible, (d) none of the above.

Q.29 In an atom, if the electron cloud is displaced from its normal position, the restoring force on the electron is proportional to -
(a) the square of the displacement,
(b) the displacement,
(c) the inverse of the displacement,
(d) none of the above.

Q.30 If a dielectric material is polarized by an alternating electric field, its electrons or ions are subjected to -
(a) displacement, (b) oscillations,
(c) annihilation, (d) none of the above.

Q.31 The frequency of vibrations inside a dielectric material with valence crystal structure, produced by its polarization by an alternating electric field, is -
(a) in ultraviolet range, (b) infrared range,
(c) low frequency range, (d) none of the above.

Q.32 The frequency of vibrations inside a dielectric material with ionic crystal structure, produced by its polarization by an alternating electric field, is -
(a) in ultraviolet range, (b) infrared range,
(c) low frequency range, (d) none of the above.

Q.33 The electric dipole moment (P) per unit volume produced in a dielectric material, by the application of a static electric field (E) on it is given by -
(a) $\varepsilon_o(\varepsilon_r - 1)$ E, (b) ε_r.E,
(c) $\varepsilon_o.\varepsilon_r$.E, (d) none of the above.

Q.34 The electric dipole moment produced in a solid dielectric material, by the application of an alternating electric field on it, is -
(a) in phase with the applied field,
(b) out of phase of the applied field,
(c) zero,
(d) none of the above.

Q.35 Solid dielectrics can be divided into three groups (i) elemental, (ii) ionic without dipoles, (iii) permanent dipolar.
(a) true, (b) false,
(c) partially true, (d) nonsense.

Q.36 In elemental solid dielectric materials, all atoms -
(a) are of same kind,
(b) can be divided into two kinds,
(c) differ with each other,
(d) none of the above.

Q.37 In elemental dielectrics, such as diamond and phosphorous, the polarization is -
(a) not possible,
(b) by relative displacement of nuclei and electron clouds,
(c) by emission of some electrons,
(d) none of the above.

Q.38 Clussius-Mossotti expression $(\varepsilon_r - 1)/(\varepsilon_r + 2) = N\alpha_e/3\varepsilon_o$, where N is the number of atoms per cubic-meter and a the electronic polarizability, is applicable to -
(a) elemental dielectrics, (b) ionic dielectrics,
(c) dipolar dielectrics, (d) none of the above.

Q.39 The effect of temperature on the elemental dielectrics is -
(a) negligible, (b) predominant,
(c) destablizing, (d) none of the above.

Q.40 Elemental solids and liquids are usually better dielectrics, in comparison with gases, because of their -
(a) higher number of atoms per meter-cube,
(b) nondisplacibility of electron cloud,
(c) structural stability,
(d) none of the above.

Q.41 In ionic dielectrics without permanent dipoles, the ionic polarisation is dependent upon -
(a) compressibility, (b) temperature,
(c) viscosity, (d) none of the above.

Q.42 The internal field, in an dielectric material, is the field in the vicinity of an atom. In gases it is usually equal to the external field but in solids and liquids, it is different because of the -

(a) compactness of atoms, (b) rarefaction of the atoms,
(c) interatom friction, (d) none of the above.

Q.43 For the dielectrics under an alternating electric field, the dielectric constant is a complex number because, under this condition, -

(a) polarizability is complex, (b) frequency is complex,
(c) applied field is complex, (d) none of the above.

Q.44 Dielectric constant under alternating fields, of idealized non-dipolar solid dielectrics remains same as static dielectric constant, for the frequencies -

(a) above infrared frequency,
(b) below infrared frequency,
(c) below ultraviolet frequency,
(d) none of the above.

Q.45 Dipolar dielectrics, under the influence of an external electric field, get polarized. When the field is removed, they come to their normal state. The time constant to come to normal state is known as -

(a) normalization time, (b) relaxation time,
(c) dipole time, (d) none of the above.

Q.46 In dipolar substances the orientational polarization is complex, which diminishes to zero at higher frequencies because -

(a) of the low relaxation time of the dipoles,
(b) the dipoles high relaxation time,
(c) presence of dipoles,
(d) none of the above.

Q.47 The energy absorbed by a dielectric from an alternating applied electric field source, known as dielectric loss is proportiona to -

(a) the real part of the complex dielectric constant.
(b) the imaginary part of the complex dielectric constant,
(c) the temperature,
(d) none of the above.

Q.48 Magnetic susceptiblity of a material is given by -

(a) 1/magnetic permeability,
(b) 1/ permitivity,
(c) (relative permeability - 1),
(d) none of the above.

Q.49 If H is the magnetic field intensity, and M the magnetization of a material, the fux density (B) in the material is given by -

(a) $B = \mu_o (M + H)$, (b) $B = M + \mu H$,
(c) $B = \mu(M + H)$, (d) none of the above.

Q.50 Apart from the original dipole of an electron, a new dipole is induced in it under the influence of a magnetic field, in a direction -

(a) enhancing the magnetic field,
(b) opposing the magnetic field,
(c) at right-angle to the magnetic field,
(d) none of the above.

Q.51 Diamagnetic materials, have -

(a) permanent magnetic dipoles,
(b) no permanent magnetic dipoles,
(c) resonant magneticdipoles,
(d) none of the above.

Q.52 In paramagnetic materials, interaction between permanent magnetic dipoles present in the material is -

(a) such that dipoles become parallel,
(b) such that dipoles become anti-parallel,
(c) zero or negligible,
(d) none of the above.

Q.53 In ferromagnetic materials, interaction between permanent magnetic dipoles present in the material is -

(a) such that dipoles become parallel,
(b) such that dipoles become anti-parallel,
(c) zero or negligible,
(d) none of the above.

Q.54 In antiferromagnetic materials, interaction between permanent magnetic dipoles present in the material is -

(a) such that dipoles become parallel,
(b) such that dipoles are equal and become anti-parallel,
(c) zero,
(d) none of the above.

Q.55 If in a material with permanent magnetic dipoles, the dipoles interact to become anti-parallel, but still some magnetization remains due to the unequal magnitudes of the adjascent dipoles, such a material is called -

(a) diamagnetic, (b) ferromagnetic,
(c) paramagnetic, (d) ferrimagnetic..

Q.56 Under the influence of a magnetic field, magnetic dipoles are induced in -

(a) diamagnetic materials, (b) ferromagnetic materials,
(c) paramagnetic materials, (d) all types of materials..

Q.57 For practical purposes, the relative magnetic permeability of all diamagnetic materials may be taken as -

(a) zero, (b) 0.5,
(c) one, (d) none of the above.

Q.58 A permanent magnetic dipole is produced whenever a charged particle has -

(a) a displacement, (b) velocity moment,
(c) angular moment, (d) none of the above.

Q.59 In atomic structure, the angular momentum are produced due to (i) electron's orbital motion, (ii) electron spin, (iii) nuclear spin.

(a) all the above are true, (b) only first two are true,
(c) only last two are true, (d) none of the above is true.

Q.60 From engineering point of view, the only angular moment of the atomic structure, of interest is that due to -

(a) electron's orbital motion, (b) electron's spin,
(c) nuclear spin, (d) none of the above.

Q.61 For practical purposes, the relative magnetic permeability of all paramagnetic materials may be taken as -

(a) zero, (b) 0.5,
(c) one, (d) none of the above.

Q.62 At temperatures above the ferromagnetic currie temperature, the ferromagnetic materials, somehow, behave like -

(a) diamagnetic materials, (b) paramagnetic materials,
(c) ferrimagnetic materials, (d) none of the above.

Q.63 The ferromagnetic materials exhibit hysterisis in the B versus H curves only -

(a) at ferromagnetic Curie temperature,
(b) below ferromagnetic Curie temperature,
(c) above ferromagnetic Curie temperature,
(d) none of the above

Q.64 Coercivity of the ferromagnetic materials, wrt. B-H loop, is -

(a) the field required to bring remanant magnetization to zero,
(b) magnetization when the applied magnetic field is removed.
(c) the field required to bring magnetic saturation,
(d) none of the above.

Q.65 In deciding on application of a ferromagnetic material for a specific purpose, the most important property to be considered is -

(a) retentivity, (b) coercivity,
(c) saturation, (d) none of the above.

Q.66 As compared to the coercivity of the pulse transformer core, the coercivity of a permanent magnet material is -

(a) higher, (b) lower,
(c) may be same, (d) none of the above.

Q.67 Retentivity of the ferromagnetic materials, wrt. B-H loop, is -

(a) the field required to bring remanant magnetization to zero,
(b) magnetization when the applied magnetic field is removed..
(c) the field required to bring magnetic saturation,
(d) none of the above.

Q.68 In ferromagnetic materials, the magnetic retentivity is because-

(a) all the atomic dipoles are aligned together,
(b) most of the atomic dipoles are aligned together,
(c) almost half the atomic dipoles are aligned together,
(d) none of the above.

Q.69 The most important characteristic regading ferromagnetism of a ferromagnetic material is -

(a) retentivity, (b) coercivity,
(c) saturation, (d) none of the above.

Q.70 In ferromagnetic materials, the internal field field, seen by a given dipole is equal to the applied field plus a contribution from the neighbouring dipoles, which tends to align it in the same direction as its neighbours. This is -

(a) ferromagnetism law, (b) Weiss hypothesis,
(c) Curie-Weiss law, (d) none of the above.

Q.71 At temperatures (T in degree kelvin) above ferromagnetic Curie temperature, the relative magnetic permeability (μr) of ferromagnetic materials depends upon temperature and is given by (μr - 1) = C /(T - t), where C is Curie constant, and t is paramagnetic Curie temperature. This is -

(a) ferromagnetism law, (b) Weiss hypothesis,
(c) Curie-Weiss law, (d) none of the above.

Q.72 Curie-Weiss law is applicable for temperatures -

(a) below Curie temperature,
(b) above paramagnetic Curie temperature,

(c) at Curie temperature,
(d) none of the above.

Q.73 In ferromagnetic materials spontaneous magnetization occurs when the material temperature is -
(a) above Curie temperature,
(b) above ferromagnetic Curie temperature,
(c) below feromagnetic Curie temperatures,
(d) none of the above.

Q.74 In antiferromagnetic and ferrimagnetic materials, the antiparallel alignment is caused due to -
(a) very low distance between the interacting atoms,
(b) very large distance between the interacting atoms,
(c) large number of free electrons,
(d) none of the above.

Q.75 In antiferromagnetic materials, the susceptibility rises to a peak value at -
(a) Neel temperature, (b) Curie temperature,
(c) absolute zero temperature, (d) none of the above.

Q.76 In antiferromagnetic materials at zero degree kelvin, the dipoles are grouped and aligned to make larger dipoles with spontaneous magnetization. As the temperature is raised the groupings get smaller and the spontaneous magnetization disappears at -
(a) Curie temperature, (b) Neel temperature,
(c) ambient temperature, (d) none of the above.

Q.77 MnF_2, MnO_2, MnO, MnS, FeO, NiF_2 and CoO are the examples of -
(a) diamagnetic materials, (b) paramagnetic materials,
(c) ferrimagnetic materials, (d) antiferromagnetic materials.

Q.78 Fe, Co, Ni, Mu-metal, Supermalloy, Alnico V, Platinum-Cobalt are the examples of -
(a) diamagnetic materials, (b) paramagnetic materials,
(c) ferromagnetic materials, (d) none of the above.

Q.79 $CrCl_3$, Cr_2O_3, $MnSO_4$, Fe2O_3, $FeCl_2$, $NiSO_4$ are the examples of -
(a) diamagnetic materials, (b) paramagnetic materials,
(c) ferrimagnetic materials, (d) none of the above.

Q.80 Al_2O_3, $BaCl_2$, NaCl, Diamond, Graphite, Cu, Au, Ge, Si, Se are the examples of -
(a) diamagnetic materials, (b) paramagnetic materials,
(c) ferrimagnetic materials, (d) none of the above.

Q.81 Ferrites are solid solutions of two ferromagnetic material oxides and possess very low conductivity. These are classed as -
(a) diamagnetic materials, (b) paramagnetic materials,
(c) ferrimagnetic materials, (d) none of the above.

Q.82 Electrical conductivity of silver is -
(a) 0.6×10^8 per ohm-meter, (b) 2×10^{-2} per ohm-meter,
(c) 0.5×10^{-10} per ohm-meter, (d) none of the above.

Q.83 Electrical conductivity of germanium is -
(a) 0.6×10^8 per ohm-meter, (b) 2×10^{-2} per ohm-meter,
(c) 0.5×10^{-10} per ohm-meter, (d) none of the above.

Q.84 Electrical conductivity of metals is attributed to -
(a) atoms having surplus electrons,
(b) valence electrons in short-supply,
(c) all valance electrons forming a cloud of free electrons,
(d) none of the above.

Q.85 Conductivity of metals -
(a) increases with increase of temperature,
(b) decrease with increase of temperature,
(c) does not depend on temperature,
(d) none of the above.

Q.86 Ohm's law can be stated as J = cE, where J is current density and E is field strength, and c is -
(a) resistivity, (b) conductivity,
(c) permitivity, (d) none of the above.

Q.87 The energy level of free electrons, at absolute zero, below which all energy levels are filled and above which empty, is known as -
(a) critical level, (b) transient level of energy,
(c) Fermi level of electrons, (d) none of the above.

Q.88 At Fermi level W_f, the velocity V_f of electrons each of mass m, is given by the equation -
(a) $W_f = m.V_f$, (b) $W_f = m. V_f^2$,
(c) $W_f = ½.m.V_f^2$, (d) none of the above.

Q.89 In metals electrons keep on colliding. For a velocity v and average time between collisions as T, the mean free path of an electron is given by -
(a) p = v.T, (b) p = v / T,
(c) $p = v^2.T$, (d) none of the above.

Q.90 Theoritically, the resistivity of pure metal crystal at zero degree absolute is -

(a) infinite, (b) zero,
(c) none of the above.

Q.91 According to classical physics, mean free path of an electron in a metal is a measure of resistivity of the metal. Longer the mean free path -

(a) higher the resistivity, (b) lower the resistivity,

Q.92 According to Matthiessen's rule, above a temperature called the Debye temperature of the metal, as the temperature increases its resistivity -

(a) decreases, (b) increases instantly,
(c) increases linearly, (d) none of the above.

Q.93 For metals, the Debye temperature is dependent of hardness. Higher the hardness -

(a) lower the Debye temperature,
(b) higher the Debye temperature.

Q.94 When a little copper is added to nickel, the resistivity of nickel -

(a) increase, (b) decreases,
(c) remains same.

Q.95 Thermal coductivity of metals above Debye temperatures is -

(a) constant,
(b) proportional to temperature,
(c) inversely proportional to temperature,
(d) none of the above.

Q.96 For metals at absolute temperature T, if electrical conductivity is c and thermal conductivity K then K/cT, is constant for -

(a) a particular metal, (b) all metals,
(c) all materials, (d) none of the above.

Q.97 For all metals at absolute temperature T, above the Debye temperatures, if electrical conductivity is c and thermal conductivity K then K/cT, is constant known as, and the value is -

(a) Debye constant, 2.45×10^{8} watt ohm degree-square,
(b) Lorenz constant, 2.45×10^{-8} watt ohm degree-square,
(c) none of the above.

Q.98 The value of expression K/cT, where K is thermal conductivity and c is electrical conductivity of a metal at absolute temperature

T, for all metals above Debye temperatures, is a constant. This is -

(a) Lorenz law, (b) Wiedmann-Franz law,
(c) Debye law, (d) none of the above.

Q.99 Semiconductors, as a class and in properties are closer to -

(a) metals, (b) insulators,
(c) magnetic materials, (d) none of the above.

Q.100 Semiconductors are characterized by -

(a) resistivity in between those of metals and insulators,
(b) positive temperature coefficient of resistivity,
(c) negative temperature coefficient of resistivity,
(d) none of the above.

Q.101 Semiconductors at zero degree absolute, behave as -

(a) perfect conductors, (b) perfect insulators,
(c) none of the above.

Q.102 Resistivity of semiconductors takes to transition as it -

(a) increases above absolute zero degree,
(b) decreases above absolute zero degree,
(c) increases below room temperatures,
(d) decreases below room temperatures.

Q.103 The conductivity of semiconductors depend upon -

(a) free electrons movement,
(b) holes movement,
(c) movement of electrons and holes in a direction,
(d) movement of holes and electrons in reveres directions..

Q.104 Number of electrons and that of holes in intrinsic semiconductors-

(a) are always equal, (b) are always different,
(c) may be different, (d) none of the above.

Q.105 Extrinsic semiconductors are made from intrinsic semiconductors, by -

(a) adding extra electrons, (b) adding extra holes,
(c) adding impurities, (d) none of the above.

Q.106 The conductivity of a semiconductors material depends upon its energy gap, which is difference in energy levels of -

(a) an electron and a hole,
(b) a free electron and a valence electron,
(c) a valence electron and a hole,
(d) none of the above.

Q.107 Addition of a small impurity to a molten pool of a semiconductor is called -

(a) isotoping, (b) biasing,
(c) doping, (d) none of the above.

Q.108 The Si and Ge, the most important of the semiconductor electronics, are doped with elements of -

(a) 4th column of the periodic table,
(b) 3rd column of the periodic table,
(c) 5th column of the periodic table,
(d) 3rd or 5th column of the periodic table.

Q.109 The doping impurity of a semiconductor is called a donor if it -

(a) has a surplus electron,
(b) lacks in the desired electron,
(c) none of the above.

Q.110 The doping impurity of a semiconductor is called an acceptor if it -

(a) has a surplus electron,
(b) lacks in the desired electron,
(c) none of the above.

Q.111 For doping of Si or Ge semiconductors, a donor is an element from -

(a) 3rd column of the periodic table,
(b) 4th column of the periodic table,
(c) 5th column of the periodic table,
(d) 4th or 5th column of the periodic table,

Q.112 For doping of Si or Ge semiconductors, an acceptor is an element from -

(a) 3rd column of the periodic table,
(b) 4th column of the periodic table,
(c) 5th column of the periodic table,
(d) 4th or 5th column of the periodic table,

Q.113 A semiconductor doped with a donor element is called -

(a) p-type semiconductors, (b) n-typo semiconductors,
(c) intrinsic semiconductor, (d) extrinsic semiconductor.

Q.114 A semiconductor doped with an acceptor element is called -

(a) p-type semiconductors, (b) n-type semiconductors,
(c) intrinsic semiconductor, (d) extrinsic semiconductor.

ABSTRACT

2.1. CLASSIFICATION OF MATERIALS

Materials used in electrical engineering applications are classified as follows:

1. Conductive materials

Materials which usually allow flow of electric current through them are called conductive materials. Resistor materials are also included in this class. Copper, aluminium, silver etc. metals are the examples.

2. Dielectric materials

Materials which obstruct flow of current through them and do not cause any appreciable power loss are kept in this class. Paper, mica, polymers, ceramics etc. are dielectric materials. Sometimes these are also called as insulators.

3. Magnetic materials

Materials which get magnetised by closeness with other magnets or electric current around them, are called magnetic materials. Usually these are solids and can be conductive or dielectric materials. Iron, cobalt, antimony are some magnetic materials.

4. Semiconductors

Materials in between the two classes - conductive and dielectric, with respect to the electric conductivity are known as semiconductors. Silicon and Germanium are most frequently used semiconductors in electronics industry.

Further to the above classification, the following points are notable :

(a) All the above classes consists of elements, as well as their compounds with other elements. Iron as well as steel are magnetic materials, copper and brass both are conductive materials.

(b) Very few materials can be put into the above classes rigidly. A material may be a conductive as well as magnetic, some compounds may be used as conductors in one application and dielectric in the other application. Air is usually a dielectric but it is used as a conductor at high temperatures in MHD generators. Some materials change to another class by conversion to other form.

(c) Alloying or compounding with other elements may change the class or some characteristic of a material.

(d) With change of temperature, material characteristics are altered. At absolute zero, most materials become dielectrics, while at high temperatures, many materials become conductive which were otherwise dielectric.

(e) Boundary lines between conductive materials and semiconductors with respect to their conductivity is hazy. For example, nichrome with a resistivity of 10^{-6} Ω.m is a conductive material while Sn(gray) with rescitivity of 2×10^{-6} is a semiconductor. Carbon is treated in both classes as conductive as well as semiconductor.

Hence a material's actual class is determined according to the conditions and the nature of applications.

Inter-atomic bonds

The behaviour of a material, as far as conductivity is concerned is dependent upon the bonds between various atoms of the material. These bonds may be put into three types :

(i) Metallic bond

Elements which have 1, 2 or 3 valence electrons easily give up their valence electrons to combine with other atoms or remain free. Since the positive and negative charges of an atom are equal, parting of some electrons from it leaves a positive surplus charge while the parted electrons form a negatively charged electron cloud. Such materials are good conductors and usually are metals. The parted electrons keep transitory bonds with all the atoms of the material, creating an electron cloud which is linked with all the atoms but not linked with any particular atom for all the time.

(ii) Covalent bond

This type of bonding is said to occur when valence electrons are contributed by both the partner atoms of a bond. Since no free electrons are available for the flow of charge, such materials possess poor conductivity forming insulators and semiconductor classes of materials.

(iii) Ionic bond

Some material crystals may have a surplus charged particle or a vacancy of a charged particle due to some irregularity in structure or an impurity, creating ions positively or negatively charged. Such ions become the originators of some conductivity in the material. Alkali halides such as NaCl are examples of ionic bonded materials. Ionic bonded materials show good conductivity in liquid and gaseous forms.

Usual resistivity range of conductive, semi-conduotive or dielectric materials are :

Conductive materials	: 10^{-8} to 10^{-6} Ωm
Semiconductors	: 10^{0} to 10^{2} Ωm
Dielectric materials	: 10^{12} to 10^{18} Ωm

2.2. CONDUCTIVE MATERIALS

Mostly metals are conductive materials with the following characteristic properties:

(i) Temperature coefficient of resistance

Metals have a positive temperature coefficient of resistance i.e. their resistinty increases with rise in temperature. On the other hand semiconductors and dielectrics have a negative temperature coefficient of resistance.

(ii) Atomic structure

Generally atomic structure of a material determines the class of material. All conductive materials are characterised by an ocean of free electrons moving from one atom to another, while electrons in dielectric materials are rigidly bound to atoms. Displacement of electrons in a material is the basic cause of conductivity.

(iii) Lattice structure

Apart from thermal excitation of free valence electrons of conductors, another reasons for conductivity has been found to be the impurities and imperfections in the lattice structures of the metals, which is a dominant factor of conductivity at low temperatures. When electrons move in a material they get scattered from their original path. This scattering is caused by the deviations in the lattice structure. Scattering causes the increased resistivity of the materials. Scattering is caused by thermal vibrations also.

(iv) Debye temperature

Conductive materials have a characteristic temperature known as Debye temperature, above which the resistivity of the material vary linearly with the temperature. This is Matthiessen's rule. Hard materials have a higher Debye temperature. Debye temperatures for some materials are given below:

Metal	Ag	Al	Au	Cu	Fe	Pb
Debye Temp (K)	215	398	180	315	420	88

(v) Lorentz number

For all conductive materials, the value of fraction $K/\sigma T$ is a fixed value of 2.45×10^{-8}, where K is the thermal conductivity, σ is the electrical conductivity and T is the temperature. This is Wiedemann Franz law. This law is also applicable above Debye temperature.

(vi) Superconductivity

Superconductivity, though a close relation of conductivity is not restricted to conductive materials only. Super conductivity occurs near absolute zero temperatures, the exact value varying for different materials. For mercury this temperature is 4.5 K while for lead it is 7.22 K.

Conductive materials are further divided into low-resistivity and high-resistivity materials.

(i) Low resistivity materials

Materials used for transfer of power from one place to another should have low voltage drop and low power dissipation. These qualities demand that the material resistivity should be low. Copper and aluminium are the most widely used materials for transfer of power.

Technically, copper is superior but economic reasons justify the use of aluminium. Except motor winding wires, aluminium has replaced copper in most engineering applications. Conductivity of all materials is measured in terms of copper conductivity taken as 100 per cent. Such a conductivity standard is known as IACS (International Annealed Copper Standard).

Conductors, in most of the electrical applications except the overhead transmission lines, are used in insulated form i.e. they are covered with an insulating layer. Where space is a constraint, such as in rotating machines, synthetic enamel coating by thermosetting process is provided to the conductor wires of strips. In other applications the insulation coating may be of paper, cotton or fiberglass etc. In cables polymer coatings are used alongwith paper and other insulating materials.

Overhead line conductors are bare conductors in solid form or standard conductor composite form. Overhead line aluminium conductors have middle strands of steel wires and the composite conductor is called ACSR (Aluminium Conductor Steel Reinforced). Apart from elemental conductors, such as copper and aluminium, copper alloys such as brasses and bronzes are also used as conductors.

(ii) High resistivity materials

In applications where voltage drop or power dissipation are technical requirements such as in control devices and electrical loading, lighting, heating etc. high resistivity conductors are used. These are in the form of resistors, heating elements and lamp filaments etc. such devices are actual power consumers.

Because of power dissipation in these materials, they are subjected to high temperatures. Hence high melting points and low temperature coefficient of resistance, mechanical strength etc. are other desirable properties of such materials.

Platinum, tungston, iron, carbon etc. are some high resistivity elements used for such purposes apart from a wide variety of alloys such as manganin, constantan, nichrome, kanthal etc.

2.3. SEMICONDUCTORS

As aforesaid, semiconductors are the materials with low conductivity. They are different from insulators because conductivity of semiconductors can be enhanced by some variation of conditions of application. Hence semiconductor provide controllability of conductivity. this controllability is obtained by various means and methods, such as :

(i) Temperature

Semiconductors have a negative temperature coefficient of resistance i.e. as the temperature rises, conductivity increases. This is caused by thermal excitation of elections in the semiconductors. Thermistors and photo cells use this type of conductivity control.

(ii) Impurities

Addition of an impurity in the semiconductor material provides it with surplus charge, positive or negative, depending upon the impurity, which becomes the cause of improved conductivity of the semiconductor materials. the process of addition of impurity to a semiconductor is known as doping. Semiconductor with a surplus positive charge is called a p-type of semiconductor while with a negative surplus charge, is called a n-type of semiconductor. These are widely used in solid state electronic devices such as diodes, transistors, thyristors, etc.

(iii) Electric field

The instantaneous electronic control of conductivity is obtained by the application of an electric field on a semiconductor, which controls the flow of electrons in the semiconductor device. Transistors, thyristors, MOSFETS etc. use electric field for control of their conductivities.

(iv) Magnetic field

Application of a magnetic field on a semiconductor also regulates the flow of electrons under the principle known as Hall's effect. Magnetrons and cyclotrons are constructed on this principle.

Silicon and Germanium

Silicon and Germanium are the two most important semiconductors having universal application in electronic devices. Both belong to the 4th column of the periodic table and hence have 4 valence electrons. The inter-atom bonding is of covalent type wherein each atom of bonding pair provides one electron, thus each atom is bonded to 4 different atoms.

A pure semiconductor is called an intrinsic semiconductor and its conductivity can only be enhanced by raising its temperature, thus freeing some electrons by thermal excitation. When some impurity is added to a semiconductor to enhance its conductivity, the material is called an extrinsic semiconductor. Most of the applications use extrinsic semiconductors.

Silicon or Germanium, when doped with an impurity of third column of the periodic table, which has 3 valence electrons, creates a vacancy of one electron in the semiconductor. Such a vacancy is normally termed as the presence of a hole, equivalent to the presence of a positive charge in the material. Thus the resultant extrinsic semiconductor is of p-type.

The same intrinsic semiconductor, an addition of an impurity from the 5th column of the periodic table, gives a n-type of semiconductor.

A p-type semiconductor slice, when joined with a slice of n-type semiconductor, forms a p-n junction, widely used to fabricate diodes, thyristors, transistors etc.

2.4. DIELECTRIC MATERIALS

Dielectric materials are used for various purposes and in various ways, from insulating the conductors to piezoelectric sensors. Such materials are characterized by a constant for a particular material, known as dielectric constant. In SI unit system, the product of dielectric constant of a material and 8.854×10^{-12} is known as the permitivity of the material, where the constant 8.854×10^{-12} is denoted by ε_o as the permitivity of free space. Thus permitivity of a material is :

$$\varepsilon = \varepsilon_o \, \varepsilon_r$$

where ε_r is the relative permitivity or the dielectric constant of the material. When an electric field E is applied to an isotropic dielectric, the electric flux density produced in the dielectric is given by :

$$D = \varepsilon_o \varepsilon_r \, E.$$

An isotropic dielectric is that which has same types of properties in all directions.

The dielectric constant of a material can be determined by measuring the capacitance of a parallel plate capacitor with and without the dielectric material in between the plates. If the capacitances in the two situations are C_o and C, the value of dielectric constant of the material is $\varepsilon_r = C/C_o$.

ε_r is a dimensionless quantity whereas ε_o has the dimension of farad per meter.

Dielectric materials behave a little differently in static and alternating electric fields.

Polarization in static field

When a static field is applied to a dielectric material, it gets polarized. The polarization means creation of distance between positive and negative charges of the atoms of the material, giving rise to a dipole moment. On the application of a homogeneous electric field, the dipole moment per unit volume in the material is given by :

$$P = \varepsilon_o (\varepsilon_r - 1) \, E .$$

For a monoatomic dielectric gas, the dipole moment per unit volume is given by :

$$P = N\alpha_e E$$

where N = number of atoms per cubic meter

a_e = electronic polarizability of the dielectric

Hence $\varepsilon_o(\varepsilon_r - 1) = N\alpha_e$

$$\alpha_e = \frac{\varepsilon_o(\varepsilon_r - 1)}{N}$$

This type of polarization is known as electronic polarization. In a polyatomic dielectric, the polarization may be caused by one or more of the following reasons :

(i) Orientational polarization (P_o)

(ii) Electronic polarization (P_e)

(iii) Ionic polarization (P_i)

When a dielectric carries permanent dipole moments in it, the applied electric field tries to align the existing dipole moments. This is orientational polarization. When a dielectric molecule contains some surplus positive or negative change, the ionization takes place giving birth to ionic dipole moments. Polarization of ionic dipole moments due to an electric field is called ionic polarization.

Thus a dielectric may have permanent dipole moments per unit volume (μ_p), ionic depole moments μ_i or the electronic dipole moments produced under the influence of an electric field, hence known as induced dipole moments per unit value (μ_{ind}).

Internal electric field

Because of the presence of the dipole moments in a dielectric an internal electric field is created in the material, given by :

$$E_i = E + (\gamma/\varepsilon_o)\,P$$

where E is the applied field

γ is the internal field constant

P is the dipole moment per unit volume

The dielectric constant for elemental dielectrics for one kind of atom such as diamond, phophorus etc. having no permanent dipole moments, dielectric constant is given by the equation

$$\frac{\varepsilon_r - 1}{\varepsilon_r + 2} = \frac{N\alpha_e}{3\,\varepsilon_o}$$

For ionic dielectrics having no permanent dipoles, dielectric constant has a static value as well as electronic value given by :

$$\varepsilon_o\,(\varepsilon_{rs} - 1)\,E = P_e + P_i$$

and $\varepsilon_o\,(\varepsilon_{re} - 1)\,E = P_e$

where ε_{rs} and ε_{re} are static and electronic dielectronic constants. The difference of the two is a measure of ionic polarization.

Piezoelectric materials

Materials with unsymmetrical lattice structure exhibit the property of reciprocity of polarization as the application of a stress, i.e. mechanical stress on the material causing a strain in it, produces a dipole moment in the material and the application of an external electric field, the material is strained. This reciprocity is not exhibited by materials having symmetrical lattice structure.

The materials which do not show reciprocity are known as electrostrictive materials while those showing reciprocity are called piezoelectric materials. Piezoelectric materials are used to convert mechanical energy to electrical energy and vice-versa.

Dielectrics in alternating fields

The dielectric constant in alternating fields is a complex quantity and frequency dependent. The imaginary part of the dielectric constant gives rise to absorption of energy by the dielectric from the external field.

Major dielectric materials

On the basis of physical appearance and chemical grouping, materials used in electrical engineering applications are listed below :

(i) *Fibrous materials*: Wood, paper, cotton yarn and textile, silk, glass fiber.

(ii) *Ceramics* : Alumina, quartz, talc, felspar, silica etc.

(iii) *Mica and mica products* : Muscovite mica, phlogopite mica, synthetic mica, glassbonded mica, mica-paper, micafolium.

(iv) *Asbestos and its products* : These have been found to be unsafe for humans, hence not recommended. In US these are banned.

(v) *Glasses* : Fused quartz or silica glass, burosilicate glass (pyrox), fiber glass.

(vi) *Rubbers* : Natural rubber, butadiene rubber, butyl rubber, neoprene, hypelon, silicon rubber.

(v) *Resins* : Shellae, wood-resins, synthetic resins, polymers such as cellulose derivatives, polyhydrocarbons, vinyl polymers, polycarbonates, polyester, formaldehyde, epoxy, silicon, polyurethane, cellulose acetate (bakelite), polyethylene, PTFE (teflon), PVC, polyanide, mylar, phenol formaldehyde, urea formaldehyde.

(vi) Enamels, adhesives and varnishes.

(vii) *Insulating liquids* : Transformer oil - mineral and synthetic.

(viii) *Insulating gases* : Air, nitrogen, sulphur hexafluoride.

Classification of dielectric materials

From application point of view, the dielectric materials are classified according to the maximum permissible temperatures allowed. The major dielectric materials with their classes are listed below :

Insulation class and max. operating temperature °C	*Materials*
Class Y - 90°C	Cotton, silk, paper, press board, wood etc. not impregnated nor oil-immersed; pre with or without plasticizer, vulcanized natural rubber etc.
Class A - 105°C	Cotton, silk, paper etc. with impregnation or immersed in a liquid dielectric such as oil, laminated wood, varnished paper, cellulose-acetate film etc.
Class E - 120°C	Synthetic resin enamels, cotton and paper laminates with formaldehyde bonding
Class B - 130°C	Mica, glass-fiber, asbestos etc. with suitable bonding substances, built-up mica products, glass-fiber and asbestos laminates
Class F - 155°C	Class B materials with thermally-resistant bonding materials
Class H - 180°C	Glass fiber, asbestos, built-up mica etc. with silicon resin binder
Class C above 180°C	Mica, ceramic, glass, quartz, asbestos with inorganic binders

2.5. MAGNETIC MATERIALS

Materials, in general, may be classified as those having permanent magnetic dipoles and those not having permanent magnetic dipoles. The second class of materials are called diamagnetic materials. Examples are alumina, diamond, graphite, NaCl, Cu, Au, Ge, Si, Se etc.

The materials having permanent magnetic dipoles are further classified.

(i) Paramagnetic materials

In paramagnetic materials, the magnetic dipoles are randomly directed and they have no mutual interaction. These materials have a small but positive relative permeability, near unity value. $CrCl_3$, Cr_2O_3, CoO, $MnSO_4$, Fe_2O_3, $Fe_2(SO_4)_3$, FeCls, $FeSO_4$ and $NiSO_4$ are some examples.

(ii) Ferromagnetic materials

In ferromagnetic materials, the dipoles tend to align in parallel with each other forming magnetic domains. These are generally crystalline solids. These may further be classified as soft and hard magnetic materials. Soft magnetic materials have a steeply rising magnetization curve, small and narrow hysterisis loop and hence low hysterisis losses. Soft iron, silicon steel, nickel iron alloys and soft ferrites are examples of this class. Electric machine cores are built-up from such materials.

Hard magnetic materials have a wide hysterisis loop and hence high hysterisis losses. These may be used in dielectromagnets but not in ac fields. Steels such as carbon steel, cobalt steel, tungsten steel, alnico are its examples. Permanent magnets are made from hard magnetic materials. Ferromagnetic materials behave like paramagnetic materials above Curie-temperatures.

(iii) Antiferromagnetic materials

In these materials the magnetic dipoles tend to align in antiparallel. The characteristic feature of an antiferro-magnetic material is the occurrence of a sharp maxima in the susceptibility-versus-temperature curve. The temperature at which the maxima occurs is known as the Neel temperature. Its role is same as that of Curie temperature in ferromagnetic materials.

MnF_2, MnO_2, MnO, MnS, Feo, N_1F_2 and CoO are some complex of antiferromagnetic materials.

(iv) Ferri magnetic materials

In these materials the dipole moments are of unequal magnitudes and they interact to align in antiparallel directions. Ferrites are ferrimagnetic materials. Below a certain temperature, they behave as ferromagnetic materials. Those have very high resistivity hence eddy currents are negligible even at very high frequencies.

3

Basic Electronics

Q.1 A BJT is a -

(a) voltage-controlled device,
(b) current-controlled device,
(c) power-controlled device,
(d) none of the above.

Q.2 An FET is a -

(a) voltage-controlled device,
(b) current-controlled device,
(c) power-controlled device,
(d) none of the above.

Q.3 A diode is a -

(a) nonlinear device,
(b) linear device,
(c) near-linear device,
(d) none of the above.

Q.4 An electron has a charge equal to -

(a) $+ 1.60 \times 10^{-19}$ C,
(b) $- 1.60 \times 10^{-19}$ C,
(c) $- 1.60 \times 10^{19}$ C,
(d) $+ 1.60 \times 10^{19}$ C.

Q.5 In a conducting metal, free electrons move -

(a) in orbits only,
(b) in the matter randomly,
(c) in metal from one boundary to the other.
(d) none of the above.

Q.6 When a voltage is applied to a conductor metal, electrons move -

(a) towards higher potential,
(b) towards lower potential,
(c) alternatively towards higher and lower potentials,
(d) none of the above.

Q.7 The unit of electron mobility constant, under the effect of an electric field, is -

(a) meters per volt-second,
(b) meters per second,

(c) meters square per volt-second,
(d) none of the above.

Q.8 The unit of free electron concentration in a metal, is -
(a) cubic meter, (b) inverse cubic meter,
(c) square meter, (d) none of the above.

Q.9 The unit of electrical conductivity, is -
(a) ohm-meter, (b) ohm per meter,
(c) mho per meter, (d) none of the above.

Q.10 If Tk is temperature in degree kelvin and Tc in degree celcius, then -
(a) $T_k = T_c + 273$, (b) $T_k = T_c - 273$,
(c) $T_k = T_c/273$, (d) none of the above.

Q.11 Good conductors have a concentration of free electrons, equal to -
(a) 10^{28} to 10^{29}, (b) 10^6 to 10^8,
(c) 10^{15} to 10^{20}, (d) none of the above.

Q.12 Insulators have a concentration of free electrons, equal to -
(a) 10^{28} to 10^{29}, (b) 10^6 to 10^8,
(c) 10^{15} to 10^{20}, (d) none of the above.

Q.13 Semiconductors have a concentration of free electrons, equal to -
(a) 10^{28} to 10^{29}, (b) 10^6 to 10^8,
(c) 10^{15} to 10^{20}, (d) none of the above.

Q.14 Electron-volt (eV) is the unit of energy, such that one eV is equal to -
(a) 1.60×10^{19} Joules, (b) 1.60×10^{-19} Joules,
(c) -1.60×10^{19} Joules, (d) none of the above.

Q.15 In metals, the current is conducted by electrons only hence these are called unipolar. In materials like silicon and germanium, the current can be conducted by holes as well as electrons, hence these are called -
(a) semiconductors, (b) bipolar,
(c) bidirectional, (d) none of the above.

Q.16 With rise in temperature, the conductivity of semiconductor materials increases, because of the increase of concentration of -
(a) free electrons, (b) holes,
(c) holes and free electrons, (d) none of the above.

Q.17 Doping is a process by which conductivity of semiconductors is increased, in doping the semiconductor is made -

(a) the purest, (b) impure,
(c) hot, (d) none of the above.

Q.18 The doped semiconductor can have either free electrons or free holes. If it has free holes, then the doped semiconductor is called -

(a) p-type, (b) n-type,
(c) p-n type, (d) none of the above.

Q.19 For making p-type semiconductors, doping is done by using -

(a) arsenic or phosphorous, (b) boron or gallium,
(c) sodium or potassium, (d) none of the above.

Q.20 If n, p, and m are respectively the free-electron, hole and intrinsic concentrations, then according to the mass-action law -

(a) $np = m^2$, (b) $n = p = m$,
(c) $n/p = m$, (d) none of the above.

Q.21 In semicoductors, if the concentration of electrons or holes is not uniformly distributed, the charge flows from the place of higher charge to that of lower charge. This process is called -

(a) doping, (b) diffusion,
(c) grading, (d) none of the above.

Q.22 In a p-n junction, the free electrons and holes near the junction boundary, combine each other and get neutralized. The boundary region of such a junction is called (i) depletion region, (ii) space-charge region, (iii) transition region.

(a) all the above are true, (b) only first two are true,
(c) only last two are true, (d) none of the above.

Q.23 In a junction diode, the flow of current is -

(a) from anode to cathode,
(b) from cathode to anode,
(c) dependent upon applied voltage,
(d) none of the above.

Q.24 In a p-n junction the processes of diffusion and drift are -

(a) analogous for electrons and holes respectively,
(b) opposite to each other,
(c) same,
(d) none of the above.

Q.25 In a p-n junction, a negative bias (i) produces a sturation current flow from cathode to anode, (ii) increases the depletion region.

(a) both are true, (b) only (i) is true,

(c) only (ii) is true, (d) none of the above is true.

Q.26 The saturation current flow, due to a reverse bias across a p-n junction is dependent upon -

(a) the biasing voltage,
(b) temperature,
(c) biasing voltage and temperature,
(d) none of the above.

Q.27 If V and I represent the voltage and current at operating point of a diode characteristic, the AC resistance of the diode is given by -

(a) V/I, (b) dV/I,
(c) dV/dI, (d) V/dI.

Q.28 When zener diode is used as a voltage regulator, the voltage-stablizing current flows in the diode, in the direction -

(a) from anode to cathode,
(b) from cathode to anode,
(c) dependent upon the applied voltage,
(d) none of the above.

Q.29 If the cathodes of two zener diodes are are joined together and an AC voltage applied to the two anodes through a high resistance, the votage wave shape at the two anodes will be -

(a) same as that of input,
(b) same as that of input with peaks clipped,
(c) half wave rectified,
(d) full wave rectified.

Q.30 For a rectifier circuit, the output ripples can be reduced by connecting a capacitor -

(a) in series with the load, (b) in parallel with the load,
(c) across the input supply, (d) none of the above.

Q.31 A a.c. voltage can provide its double-valued d.c voltage by the use of -

(a) two rectifier circuits,
(b) one clamping and one rectifying circuit,
(c) through transformer only,
(d) none of the above

Q.32 In semiconductor devices, the covalent bonds of semiconductors are loosened by -

(a) electric field, (b) temperature,
(c) magnetic field, (d) none of the above.

Q.33 Temperature coefficients of metals and semiconductors resistances are -

(a) positive,

(b) negative,

(c) positive and negative respectively,

(d) none of the above.

Q.34 Out of the three terminals of a transistor - base, emitter and collector, the base is -

(a) one end of pnp junction, (b) middle of the pnp junction,

(c) the junction itself, (d) none of the above.

Q.35 For a BJT, common-base and common-emitter large-signal current gains are respectively in the range -

(a) 0.95 - 0.998 and 50 - 200, (b) 50 - 200 and 0.95 - 0.998,

(c) 20 - 50 for both, (d) none of the above.

Q.36 A BJT is said to be in active mode when its B-E and B-C junctions are respectively -

(a) forward and reverse biased, (b) reverse and forward biased,

(c) reverse biased both, (d) forward biased both.

Q.37 A BJT is in cutoff state when its B-E and B-C junctions are respectively -

(a) forward and reverse biased, (b) reverse and forward biased,

(c) reverse biased both, (d) forward biased both.

Q.38 A BJT is in saturation (ON) state when its B-E and B-C junctions are respectively -

(a) forward and reverse biased, (b) reverse and forward biased,

(c) reverse biased both, (d) forward biased both..

Q.39 The common-base collector characteristics of a BJT are plots of -

(a) C-current versus C-B voltage,

(b) C-current versus C-B voltage for E-current variation,

(c) C-current versus C-E voltage,

(d) C-current versus C-E voltage for B-current variation.

Q.40 The common-base collector characteristics of a BJT are plots of -

(a) C-current versus C-B voltage,

(b) C-current versus C-B voltage for E-current variation,

(c) C-current versus C-E voltage,

(d) C-current versus C-E voltage for B-current variation..

Q.41 The dc current gain of a BJT is the ratio of -
(a) collector current to base current,
(b) collector current to emitter current,
(c) emitter current to base current,
(d) none of the above.

Q.42 The turn-on time of a BJT is the sum of -
(a) delay time and rise time,
(b) delay time and storage time,
(c) rise time and fall time,
(d) none of the above.

Q.43 The turn-off time of a BJT is the sum of -
(a) delay time and rise time,
(b) storage time and fall time,
(c) storage time and fall time,
(d) none of the above.

Q.44 The current gain ofa BJT is approximately the ratio of -
(a) collector current and emitter current,
(b) collector current and base current,
(c) emitter current and base current,
(d) none of the above.

Q.45 Noise margin of a BJT is the minimum noise voltage at input that causes the output to deviate from its normal value.
(a) true, (b) false,
(c) partially true.

Q.46 The propagation delay of a BJT is -
(a) sum of turn-on and turn-off delays,
(b) average of turn-on and turn-off delays,
(c) difference of turn-off and turn-on delays,
(d) none of the above.

Q.47 In a Schottky diode, the junction is made-up of -
(a) p and n semiconductors,
(b) metal and n semiconductor,
(c) none of the above.

Q.48 The switching speed of a Scottky diode, as compared to that of a pn junction diode is -
(a) higher, (b) lower,
(c) same,

Q.49 A JEFT is a -
(a) unipolar device, (b) bipolar device,
(c) tripolar device, (d) none of the above.

Q.50 In an n-channel JEFT, there are two p-gates on an n-channel. The three leads - one gate, one source and one drain, we get in the device, are connected -

(a) one each to p, n, p semiconductor,
(b) n is gate, one p is source and the other p is drain,
(c) both source and drain are from n-channel and gate is from both p-semiconductors,
(d) none of the above.

Q.51 In a JEFT, the minimum voltage required between gate and source, to completely deplete the channel, is called -

(a) bias voltage, (b) pinchoff voltage,
(c) gate-source bias, (d) none of the above.

Q.52 In JEFT, both source and drain are connected to the common channel, the voltage applied between the two leads is not shorted because of -

(a) the channel normal resistance,
(b) the depletion of the channel,
(c) the biasing,
(d) none of the above.

Q.53 Drain characteristics of an FET are plots of drain current versus drain-source voltage for -

(a) constant gate-source voltage,
(b) variation of gate-source voltage,
(c) variation of gate current,
(d) none of the above.

Q.54 The gate of an FET is -

(a) always zero, (b) very low,
(c) fixed, (d) none of the above.

Q.55 For an FET cutoff state, its drain current is -

(a) zero, (b) fixed,
(c) very low, (d) none of the above.

Q.56 For JEFT to be active, apart from drain current being nonzero, its channel should be -

(a) undepleted, (b) partially depleted,
(c) fully depleted, (d) none of the above.

Q.57 In a JEFT, when the channel is partially depleted, it is called to be in -

(a) active mode, (b) ohmic mode,
(c) pinched off, (d) none of the above.

Q.58 For small values of drain-source voltage, the FET behaves like -

(a) an inductor, (b) a capacitor,
(c) linear resistor, (d) none of the above.

Q.59 The drain current versus gate-source voltage characteristics of an FET are called -

(a) amplification characteristics, (b) transfer characteristics,
(c) control curves, (d) none of the above.

Q.60 In a MOSFET, also known as an insulated-gate FET, the gate is insulated from the channel of the device by a layer of -

(a) any insulation, (b) silicon dioxide,
(c) metal, (d) none of the above.

Q.61 The resistance between the gate and the source terminal of a MOSFET is of the range -

(a) E5 - E8 ohms, (b) E10 - E15 ohms,
(c) E25 - E30 ohms, (d) none of the above

Q.62 The gate current of a MOSFET is always -

(a) constant,
(b) dependent upon applied voltage,
(c) zero,
(d) none of the above.

Q.63 In a depletion mode MOSFET, the path of current between the source and drain terminals is provided by the depletion of the channel in between. The depletion is caused by -

(a) source-drain applied voltage,
(b) negative gate-source bias,
(c) gate current,
(d) none of the above.

Q.64 In an enhanced mode MOSFET, the channel between the source and drain is not depleted but further widened. This widening of the channel is caused by -

(a) positive gate-source bias,
(b) source-drain applied voltage,
(c) gate current,
(d) none of the above.

Q.65 In a MOSFET, gate-source bias voltage -

(a) is always positive,
(b) is always negative,
(c) can be positive or negative,
(d) none of the above.

Q.66 On an IC chip, as compared to the space required by a transistor, the space required by a rèsistor is -

(a) almost same, (b) much larger,
(c) much less, (d) none of the above.

Q.67 In integrated chips, transistors can be used in place of resistors, it saves space.

(a) true, (b) false,
(c) nonsense.

Q.68 In an enhancement MOSFET, no channel between the source and drain is necessary. It is made in the substrate by -

(a) a positive bias of the gate-source terminals,
(b) a negative bias of the gate-source terminals,
(c) an applied voltage between the drain and source,
(d) none of the above.

Q.69 The minimum voltage required to bias the gate-source junction of an Enhancement-MOSFET, so that drain source current can be established, is known as -

(a) gate-source critical voltage,
(b) gate-source threshold voltage,
(c) conduction bias,
(d) none of the above.

Q.70 A transmission gate is formed of a pair of MOSFETs, known as CMOS. CMOS is thus made up of -

(a) NMOSFET-PMOSFET combination,
(b) two NMOSFETs,
(c) two PMOSFETs,
(d) none of the above.

Q.71 Currently the most popular technology used for digital logics is -

(a) DTL, (b) TTL,
(c) FET, (d) MOSFET,

Q.72 An electronic circuitary aimed at converting a weak signal into a useful signal, is usually termed as -

(a) oscillator, (b) amplifier,
(c) amplidyne, (d) none of the above.

Q.73 The input output ratio of an amplifier is termed as -

(a) gain, (b) amplification factor,
(c) input/output ratio, (d) none of the above.

Q.74 The frequency range in which an amplifier remains affective depends upon the following component used in the amplifier circuit -

(a) resistor, (b) inductor,
(c) capacitor, (d) transistor.

Q.75 The ac current gain of a transistor is the ratio of -

(a) variation in collector current to that of the base current,
(b) collector current to emitter current,
(c) emitter current to base current,
(d) none of the above.

Q.76 For feeding ac signals to a biased transistor base, a capacitor is used in series, because it provides -

(a) low impedance to ac signal,
(b) high impedance to dc bias voltage,
(c) low impedance to ac and high impedance to dc,
(d) none of the above.

Q.77 The ac voltage gain of a transistor is the ratio of -

(a) collector voltage and base voltage,
(b) collector voltage and emitter voltage,
(c) emitter voltage and base voltage,
(d) none of the above.

Q.78 Common collector amplifier and emitter follower amplifier mean -

(a) the same thing, (b) slightly different things,
(c) quite different things.

Q.79 For a transistor CE circuit, the short-circuit input resistance is the ratio -

(a) base-emitter voltage to base current,
(b) base-emitter voltage to collector-emitter voltage,
(c) collector current to base current,
(d) collector current to collector-emitter voltage.

Q.80 For a transistor CE circuit, the open-circuit reverse voltage gain is the ratio -

(a) base-emitter voltage to base current,
(b) base-emitter voltage to collector-emitter voltage,
(c) collector current to base current,
(d) collector current to collector-emitter voltage.

Q.81 For a transistor CE circuit, the short-circuit forward current gain is the ratio -

(a) base-emitter voltage to base current,

(b) base-emitter voltage to collector-emitter voltage,
(c) collector current to base current,
(d) collector current to collector-emitter voltage.

Q.82 For a BJ transistor CE circuit, the open-circuit output conductance is the ratio -
(a) base-emitter voltage to base current,
(b) base-emitter voltage to collector-emitter voltage,
(c) collector current to base current,
(d) collector current to collector-emitter voltage..

Q.83 Class-A operation of a transistor amplifier, means that the transitor operates -
(a) in fully active (linear) region,
(b) for halfwave in active and halfwave in cutoff region,
(c) in mostly cutoff region,
(d) none of the above.

Q.84 Class-B operation of a transistor amplifier, means that the transitor operates -
(a) in fully active (linear) region,
(b) for halfwave in active and halfwave in cutoff region,
(c) in mostly cutoff region,
(d) none of the above.

Q.85 The efficiency of an amplifier is the ratio of -
(a) output signal power to the base signal power,
(b) output signal power to dc power delivered by supply,
(c) output power to input power,
(d) none of the above.

Q.86 Class-AB operation of a transistor amplifier, means that the transitor operates -
(a) in fully active (linear) region,
(b) for halfwave in active and halfwave in cutoff region,
(c) in mostly cutoff region,
(d) none of the above..

Q.87 In an ordinary FET/MOSFET, the source and the drain can be interchanged without any effect on their performance.
(a) true, (b) false,
(c) partially true.

Q.88 The usual efficiencies of class-A and B amplifiers are respectively around -
(a) 75 % and 25 %, (b) 25 % and 75 %,
(c) 50 % both, (d) none of the above.

Q.89 In a push-pull amplifier, the bias is at cut-off of the two transistors, which are -

(a) identical to each other, (b) complimentary to each other,
(c) independent of each other, (d) none of the above.

Q.90 An operational amplifier (opamp) (i) is an anolog device, (ii) has two inputs, one inverting and the other noninverting, (iii) has only one output, (iv) usualyy has four blocks - differential amplifier, high-gain amplifier, buffer/level shifter and driver.

(a) all the above are true, (b) only last three are true,
(c) only last two are true, (d) only last one is true.

Q.91 The output offset voltage of an opamp is the output voltage when -

(a) both inputs are zero,
(b) one input is zero and the other is unity,
(c) both inputs are equal and opposite,
(d) none of the above.

Q.92 The input offset voltage of an opamp is the input voltage necessary to -

(a) bring the output equal to the input,
(b) compensate for losses in the opamp,
(c) bring the output to zero,
(d) none of the above.

Q.93 The slew rate of an opamp is the maximum rate of change of output voltage when the input is -

(a) an unit impulse, (b) a unit step,
(c) a unit ramp, (d) a large step..

Q.94 For a feedback amplifier, gain C = A/(1 + AB) where A and B are respectively the feedforward and feedback path gains. The feedback is termed as positive when -

(a) C = A, (b) C > A,
(c) C < A.

Q.95 In an electronic circuit, if on total withdrawal of an input signal if the feedback is able to maintain the output at the same level, the circuit is known as -

(a) an amplifier, (b) an oscillator,
(c) a feedback circuit, (d) none of the above.

Q.96 For a feedback circuit, gain C = A/(1 + AB) where A and B are respectively the feedforward and feedback path gains. The circuit will behave like an oscillator, if -

(a) AB = -1, (b) AB = 1,
(c) A/B = 1, (d) none of the above.

Q.97 An ideal opamp has (i) infinite gain, (ii) infinite input resistance, (iii) zero output resistance.
(a) all the above are true, (b) only first two are true,
(c) only last two are true, (d) none of the above is true.

Q.98 741 is the number of an industry-standard -
(a) amplifier, (b) oscillator,
(c) opamp, (d) none of the above.

Q.99 By negative feedback in an amplifier -
(a) gain is reduced and stability is improved,
(b) gain is improved and stability is reduced,
(c) both gain and stability are diminished,
(d) both gain and stability are improved.

Q.100 For a feedback amplifier, the product of gain and bandwidth -
(a) varies with input, (b) is fixed,
(c) varies with frequency, (d) none of the above.

Q.101 In amplitude modulation, the envelop of the modulated wave is formed by the -
(a) carrier wave, (b) signal wave,
(c) carrier or signal wave, (d) none of the above.

Q.102 Resonanance in electrical circuits is -
(a) oscillation response of the circuit,
(b) current response of the circuit,
(c) voltage response of the ciruit,
(d) frequency response of the circuit.

Q.103 At resonance, the impedance of the circuit is -
(a) maximum, (b) minimum,
(c) real, (d) none of the above.

Q.104 Laplace transform is a mathematical tool to convert -
(a) functions of frequency to functions of time,
(b) functions of time to functions of frequency,
(c) differential equations to linear equations,
(d) none of the above.

Q.105 If M is the maximum value of amplitude frequency-response of an electrical system, then that frequency is known as cutoff frequency, at which the magnitude of amplitude will be -
(a) M/2, (b) $M/\sqrt{2}$,
(c) 2M, (d) none of the above.

Q.106 In an electrical system, as compared to the maximum power delivered, the power delivered at cutoff frequency is -

(a) equal, (b) half,
(c) quarter, (d) none of the above.

Q.107 In a high-pass filter -

(a) high voltage is passed through,
(b) low voltage is passed through,
(c) high frequency waves are passed through,
(d) none of the above.

Q.108 A decibel is -

(a) one-tenth of bel, (b) ten-times of bel,
(c) same as bel, (d) none of the above.

Q.109 In electrical systems Bel is a unit of -

(a) power, (b) energy,
(c) gain, (d) amplitude.

Q.110 If P_1 and P_2 are respectively the input and output powers of an electrical network, its gain, expressed in decibels, is -

(a) $\log (P_2/P_1)$, (b) $10 \log (P_2/P_1)$,
(c) $(1/10) \log (P_2/P_1)$, (d) none of the above.

ABSTRACT

The devices operating on the basis of electron flow and control are known as electronic devices. The science and technology dealing with the electronic devices is known as Electronics.

Electronic devices and circuits use electronic components which may be classified as passive and active components. Resistors, inductors and capacitors are passive components while electronic tubes and semiconductor devices are known as active components.

Resistors used in electronic circuits are of miniature size usually of 1/4 to 2W capacity. These are made of carbon film or moulding encapsulated in a dielectric material. These being of small size and a wide variety of ohmic values, a colour code system is used for their marking and recognition.

The resistor body is encircled with 4 coloured strips, indications of respectively, first significant digit, second significant digit, exponent of 10 as multiplier and lastly the tolerance. Hence if the first three strip colours indicate respectively a, b and c, then the value of resistor will be $ab \times 10^c$.

The various colours of strips stand for the following :

Colour	*Digit*	*Multiplier*	*Tolerance*
Black	0	10^0	
Brown	1	10	
Red	2	10^2	
Orange	3	10^3	
Yellow	4	10^4	
Green	5	10^5	
Blue	6	10^6	
Violet	7	10^7	
Gray	8	10^8	
White	9	10^9	
Gold	-	-	± 5%
Silver	-	-	± 10%
No strip	-	-	± 20%

Active components

Except for very high power application, electron tubes have become obsolete.

Semiconductor devices which are of value today are based upon p-n junctions.

P-N junction

We know that a p-type semiconductor has free positive charges or holes and an n-type semiconductor material has free electrons or negative charges. By joining the two types of materials with continuity of the crystal structure at the boundary, we get a p-n junction.

The charge flow in such a junction is possible from p-type material to n-type material only. Hence if a dc voltage is connected across the junction with positive pole at p-type material, a heavy current will flow. If the voltage supply connections are reversed, the current flow will be minimal. Thus the p-n junction is a path of almost short-circuit in one direction and almost open circuit in another direction, ideally.

At the junction point some electrons of n-type and some holes of p-type material recombine and provide a buffer zone in between. This is called depletion layer and works as a barrier, requiring a minimum magnitude of voltage in the forward direction below which the junction will not conduct. This minimum voltage is 0.7 volt for a silicon junction and 0.3 V for a germanium junction.

3.1. SOLID-STATE DEVICES

(i) Diode

It is a simple p-n junction with leads attached to the p- and n-type materials. The voltage current characteristic of a diode is shown in Fig. 3.1(a) with V_b as the barrier voltage and V_2 as the reverse breakdown voltage. Diodes are used to rectify ac to obtain dc. The simplest rectification circuit is the half-wave shown in Fig. 3.1(b). This circuit allows only positive half of the input wave to be present at the output.

For an output wave $v = V_m \cos \omega t$

Input voltage (rms) $V_i = V_m/\sqrt{2}$

Output voltage (dc) $= V_o = V_m/\pi$

Peak-inverse voltage $= V_m$

assuming there is no volt drop in the diode. In half-wave rectification the output voltage is low and the ripple factor is high.

For full-wave rectification two types of circuits are used - center tap full-wave rectifier shown in Fig. 3.2(a) and the bridge-rectifier circuit shown in

Fig. 3.2(b). In full-wave rectification both halves of the input wave are passed on to the output.

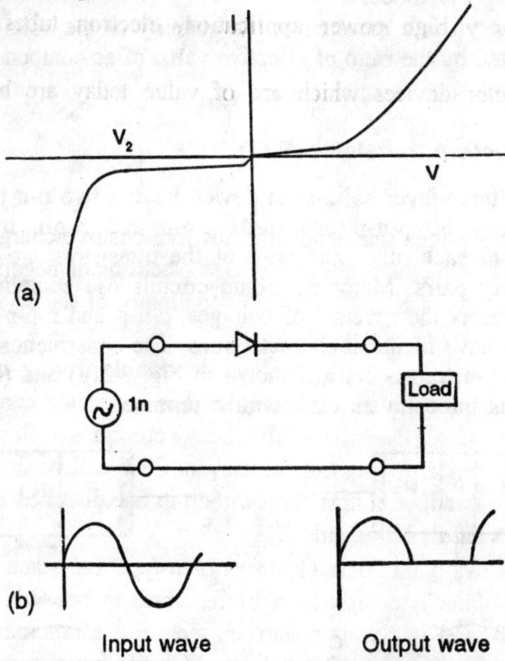

Fig. 3.1. (a) V-I characteristic of diode; (b) Half-wave rectifier.

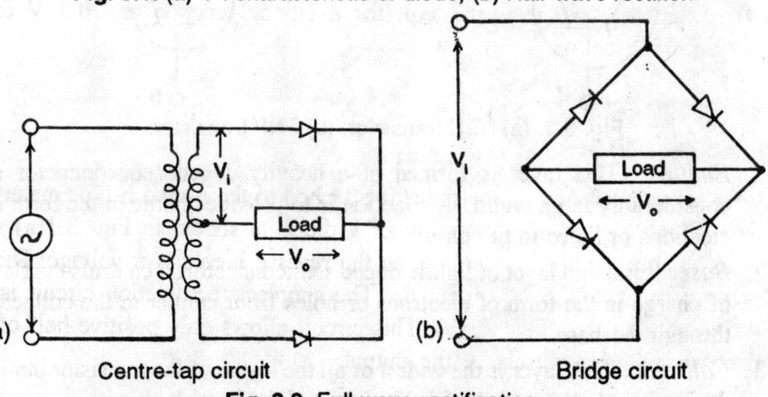

Fig. 3.2. Full wave rectification.

For the transformer tapping voltage peak V_m, the output voltage is $2V_m/\pi$. In case of bridge rectifier, for input voltage of peak V_m, the output voltage is $2V_m/\pi$. The ripple factor obtained in a full-wave rectification is 81%. The peak inverse voltage in this case is $2\,V_m$.

The two terms - peak-inverse voltage and the ripple factor are defined as:

Peak-inverse voltage (PIV) : It is the voltage peak value to which each diode of the circuit will be subjected to when reverse-biased. This rating is used in selecting the diode.

Ripple factor : It is a measure of the quantity of the voltage wave form and is represented by the ratio of effective value of ac component to average or dc component.

(ii) Bipolar junction transistor (BJT)

Transistor is a three-layer solid-state device having two p-n junctions. The three layers may be p-n-p or n-p-n. p-n-p and n-p-n transistors are complimentary to each other and most of the transistors are manufactured in complimentary pairs. Many electronic circuits use complimentary pairs of transistor. Except the reversal of voltages, p-n-p and n-p-n transistors of the same rating have identical characteristics. The construction and symbols of p-n-p and n-p-n transistors are shown in Fig. 3.3(a) and (b). Each type of transistor has three layers each with a terminal.

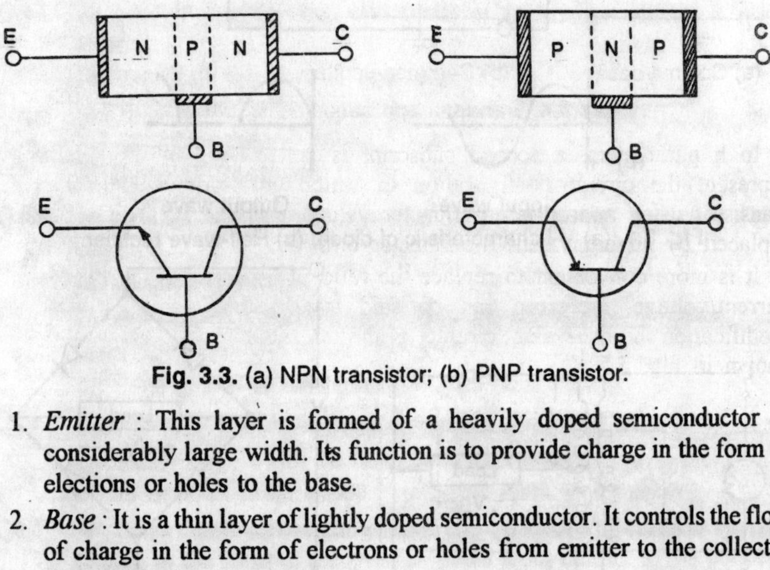

Fig. 3.3. (a) NPN transistor; (b) PNP transistor.

1. *Emitter* : This layer is formed of a heavily doped semiconductor of considerably large width. Its function is to provide charge in the form of elections or holes to the base.
2. *Base* : It is a thin layer of lightly doped semiconductor. It controls the flow of charge in the form of electrons or holes from emitter to the collector through the base.
3. *Collector* : This layer is the widest of all the layers of the transistor but its doping is in between that of the emitter and the base. It attracts change in the form of holes or electrons, arrived in the base from the emitter.

Transistor circuits are commonly represented as two-port-networks using hybrid parameters given by the general equations :

$$V_1 = h_i I_1 + h_r V_2$$

$$I_2 = h_f I_1 + h_o V_2$$

where parameters h_i is an impedance, h_r is voltage transfer junction, h_g is a current transfer junction and h_o an admittance.

Equivalent circuit

A transistor has three terminals while the h-parameter network has 4 terminals. Hence to represent a transistor circuit as equivalent h-parameter network, one of the transistor elements has to be made common to both pairs of terminals. Due to this three possibilities of circuit configurations evolve, namely - Common Base, Common Emitter and Common Collector, as shown in Fig. 3.4.

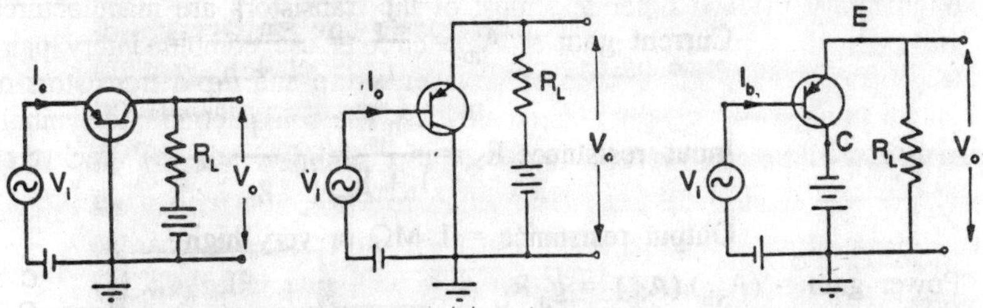

(a) Column base (b) Common emitter (c) Common collector

Fig. 3.4. Transistor application configurations.

In h parameters, a second subscript is added viz. h_{fc}, h_{ib} and h_{oc} to represent the circuit configuration to which the parameter belong. The transistor using appropriate parameter values, the transistor circuit can be replaced by an equivalent h-parameter circuit.

It is more convenient to replace the ratio of h-parameters h_f/h_i, which is current/voltage, therefore can be said transcanductance g_m. With this modification the equivalent circuits of the three transistor configurations are shown in Fig. 3.5.

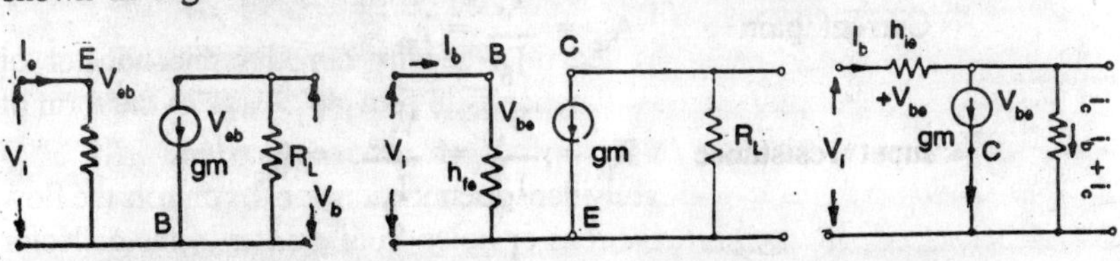

(a) Column base (b) Common emitter (c) Common collector

Fig. 3.5. Transconductance model equivalent circuits of three circuits of Fig. 3.4.

(i) Common Base Amplifier

In the circuit, the input is applied between emitter and base giving an input current I_e. Here the transenductance :

$$g_m = \frac{h_{fe}}{h_{ie}}$$

the negative sign arises because we are using voltage V_{eb} instead of V_{be} as in other circuits. From the circuit, we have :

$$I_c = -g_m V_{eb}$$

and $V_o = -I_c R_L = g_m R_L V_{eb}$

since $V_{eb} = V_i$

For CB circuit Voltage gain $A_{vb} = \frac{V_o}{V_i} = g_m R_L$

$$\text{Current gain} \quad A_{ib} = \frac{I_c}{I_b} = \frac{g_m h_{ie}}{1 + h_{fe}} = -1$$

$$\text{Input resistance } R_{ib} = \frac{h_{ie}}{1 + h_{fe}} = \frac{1}{g_m}$$

Output resistance = 1 MΩ or very high

Power gain = $(A_{vb}).(A_{ib}) = g_m R_L$

The major advantage of the circuit is its impedance transferring property, by which a low impedance source can be power-matched to a high impedance load.

(ii) Common-emitter amplifier

$$V_o = -I_c R_L = -g_m V_{be} R_L = -gm R_L V_i$$

Therefore Voltage gain $A_{ve} = \frac{V_o}{V_i} = -g_m R_L$

$$\text{Current gain} \quad A_{ie} = \frac{I_c}{I_b} = h_{fe}$$

$$\text{Input resistance} \quad R_{ie} = \frac{V_i}{I_b} = \frac{V_{be}}{I_b} = h_{ie}$$

$$\text{Output resistance} \quad R_{oe} = \frac{1}{h_{oe}} = \infty$$

$$\text{Power gain} = \frac{h_{fe}^2 R_L}{h_{ie}} \quad \text{(very high)}$$

(iii) Common-collector amplifier

Voltage gain $A_{vc} = 1$

Current gain $A_{ic} = -(1 + h_{fe}) = -h_{fe}$

For usual values of $h_{ie} = 900\ \Omega$, $h_{fe} = 24$

h_{re} = negligible h_{ve} = negligible

$A_{vc} = 1$, $R_{vc} = 49{,}000W$

Power gain = 23.5, $R_{oc} = 620\Omega$.

The output voltage follows the input voltage, hence the circuit is known as emitter follower.

Summary

The relative values of the transistor application parameters for the three configurations are given below:

Parameter	*CB*	*CE*	*CC*
Input dynamic resistance	36Ω	900Ω	49,000Ω
Output dynamic resistance	1.56MΩ	62,500Ω	620Ω
Current gain	−0.96	24	−24
Voltage gain	53.4	−53.4	0.98
Power gain	51.2	1,281	23.5

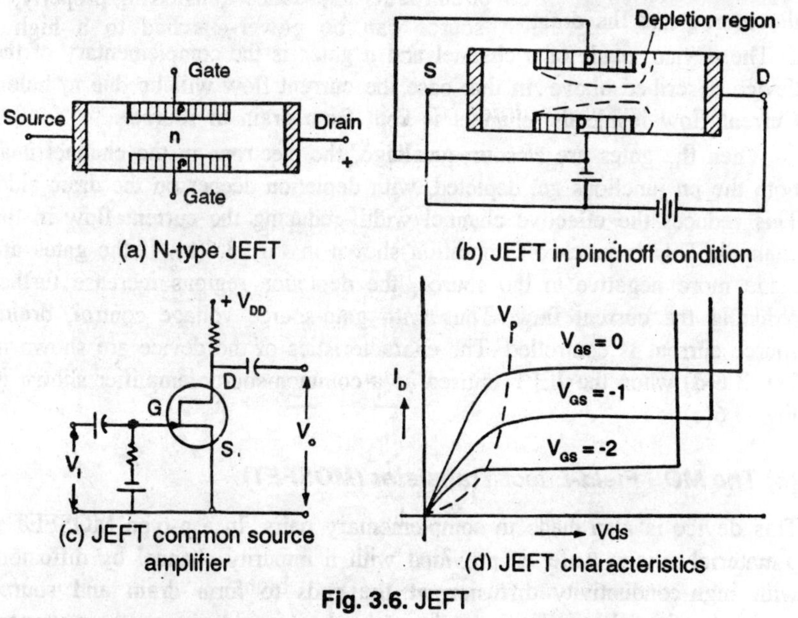

(a) N-type JEFT (b) JEFT in pinchoff condition

(c) JEFT common source amplifier (d) JEFT characteristics

Fig. 3.6. JEFT

(iii) Field-effect transistor (FET)

The field-effect transistor controls the current by application of an electric field across the conducting region. In a BJT both holes and electrons are used to conduct current, hence that device is a bipolar device. The FET uses only one type of charge for current flow, hence it is said to be unipolar.

The advantages of FET are its high input-resistance of the order of many megohms, its non-susceptibility to random variation of current or electrical noise. Hence it is useful for small input signals.

FET's of two basic designs are commonly used - junction FET, in which the gate is isolated from the conduction channel by a reverse-biased junctions, and the insulate-gate FET in which the gete is insulated from the conduction channel by a metal-oxide layer, hence also known as MOSFET (Metal-oxide Semiconductor Field Effect Transistor). The MOSFET design permits the elements to be of very small size, hence widely used in large-scale integration of switching circuits.

(a) Junction field-effect transistor

Fig. 3.6(a) shows the basic construction. A bar of n-type of semiconductor has two terminals at the ends, known as source and the drain terminals. The bar material of n-type is highly doped. On both sides of the bar two gate terminals of heavily doped and material form pn junction with the bar. The space between the two gates is called the channel. The conduction will be, in this case by the movement of electrons when a voltage is applied across the source and the drain.

The device, made of p channel and n gates is the complementary of the device described above. In that case the current flow will be due to holes. Current flow inside the channel is kept from drain to source.

When the gates are at source voltage, the electrons in the channel near both the pn junctions get depleted, with depletion deeper on the drain side. This reduces the effective channel width reducing the current flow in the channel. This is pinch off condition shown in Fig. 3.6(b). If the gates are made more negative to the source, the depletion regions increase further reducing the current flow. Thus with gate-source voltage control, drain-source current is controlled. The characteristics of the device are shown in Fig. 3.6(d) when the JEFT is used as a common-source amplifier shown in Fig. 3.6(c).

(b) The MOS Field-Effect Transistor (MOSFET)

This device is also made in complementary pairs. In a p-type MOSFET a p-material silicon wafer is provided with n impurity channel by diffusion, with high-conductivity diffusion at the ends to form drain and source terminals of the dence. The gate of metal is kept insulated from the p channel by a layer of nearly 1 micron of a dioxide, such as SiO_2. The basic construction is shown in Fig. 3.7(a).

When a negative voltage is applied to the insulated gate the n-channel between the gate and the source gets deleted, increasing the resistance between the drain and the source. If the inculated gate is applied a positive voltage, the n-channel conductivity gets enhanced. Accordingly the MOSFET can be used either in depletion or enhanced mode. The characteristics is as shown in Fig. 3.7(b).

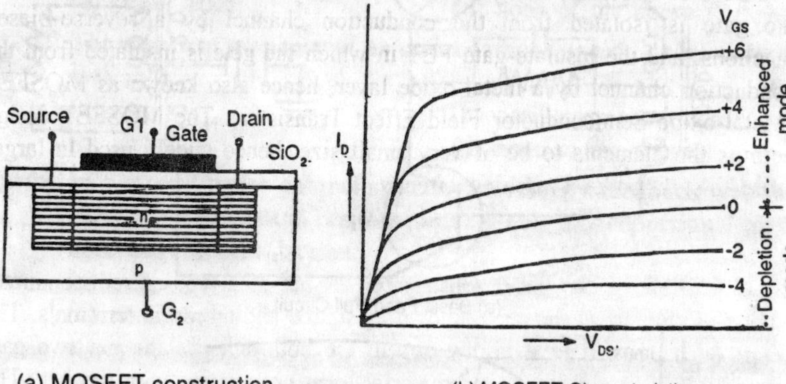

(a) MOSFET construction (b) MOSFET Characteristics

Fig. 3.7. MOSFET

3.2. TRANSISTOR APPLICATIONS

Apart from being used in amplifier circuits, transistors have many other applications too. Some important applications are given here.

(i) Push-pull circuit

Push-pull circuits are used as oscillators, amplifiers, switching circuits, frequency doubler, phase inverter etc. Every push-pull circuit uses identical pair of active devices, alternately in operation and one driving the other. Fig. 3.8(a) shows the basic circuit and (b) the operating characteristics.

The input voltage is alternately introduced to Q_1 and Q_2 in phase opposition by a centre tap transformer. The transistor currents I_1 and I_2 are in phase opposition and the load current is the difference of the two. Hence all even harmonics are eliminated from the load current. The composite load line is given by :

$$R' = \frac{2V_{cc}}{2I_{max}}$$

and the power output is given by

$$P_o = \frac{V_{cc}^2}{2R} = \left(\frac{N_2}{2N_1}\right)^2 \frac{V_{cc}^2}{2R_g}$$

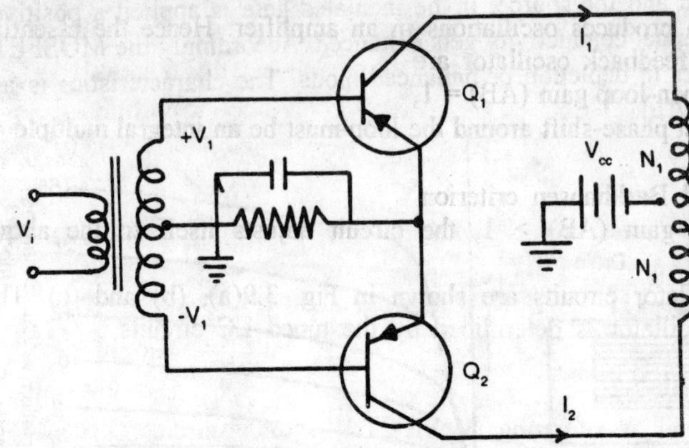

(a) Basic Pash-Pull Circuit

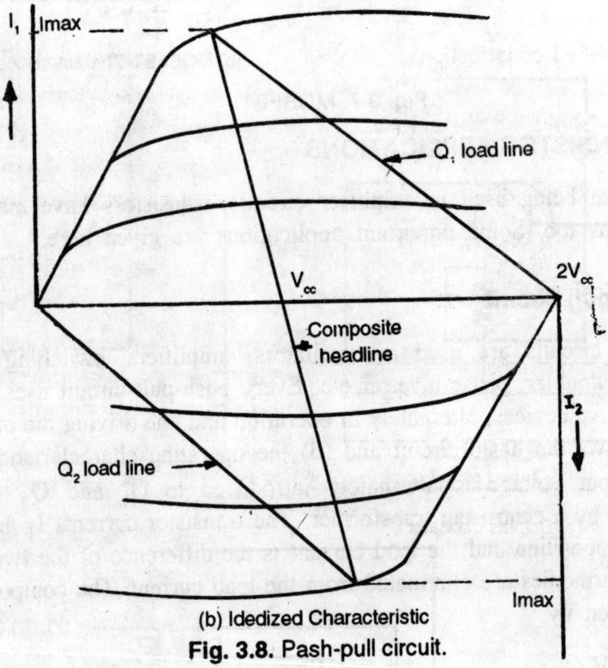

(b) Idedized Characteristic

Fig. 3.8. Pash-pull circuit.

(ii) Oscillator circuit

If the output of an amplifier is made as its input with gains of A for amplifier and B for feedback, we got the relationship

$$V_i = BV_o = -AB\ V_i$$

or $(1 + AB)\ V_i = 0$

Since output and input voltages exist,

$$1 + AB = 0$$

or $AB = -1 + j0$

This situation produces oscillations in an amplifier. Hence the essential conditions of a feedback oscillator are :

(i) The open-loop gain (AB) = 1,

(ii) The net phase-shift around the loop must be an integral multiple of 2π.

This is called Barkhausen criterion.

If the initial gain (AB) > 1, the circuit adjusts itself to the above conditions.

Various oscillator circuits are shown in Fig. 3.9(a), (b) and (c). The frequency of oscillator is determined by the tuned LC circuits.

(iii) Logic circuits

The current trend is of using MOSFETs in logic circuits because of possibility of very large-scale integration in a compact size. Fig. 3.9 shows

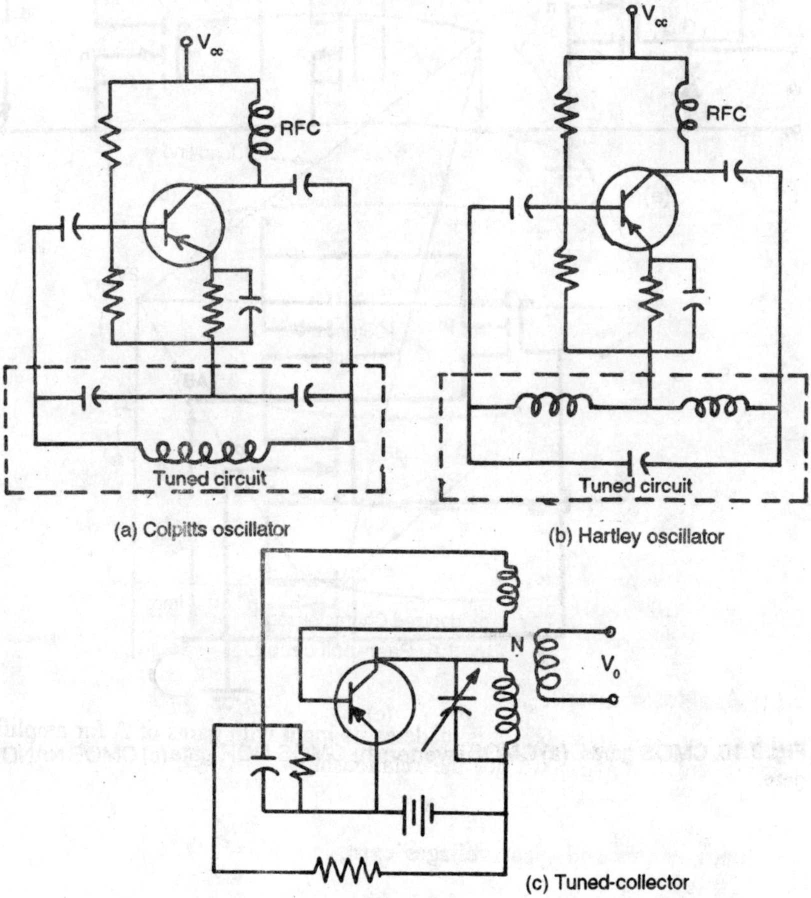

(a) Colpitts oscillator

(b) Hartley oscillator

(c) Tuned-collector Hartley oscillator

Fig. 3.9.

three gate circuits of inverter, NOR and NAND operations, using MOS devices in complementary pairs. When complementary pairs are integrated together, the device is known as a CMOS (complimentary MOS) device.

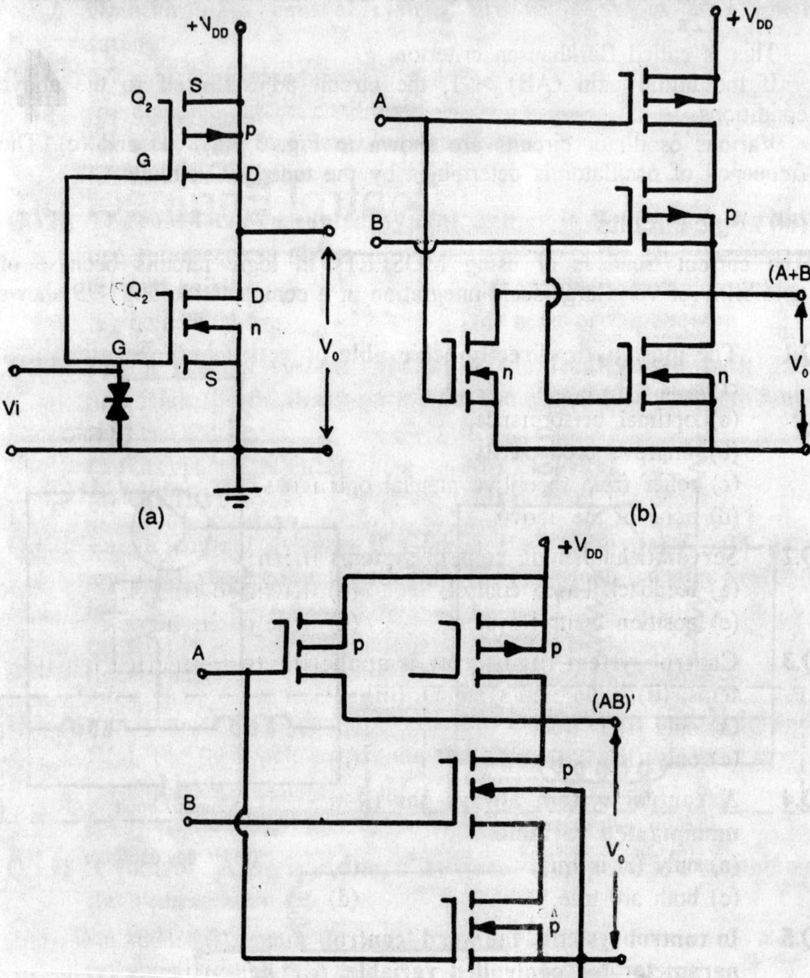

Fig. 3.10. CMOS gates. (a) CMOS inverter; (b) CMOS NOR gate (c) CMOS NAND gate.

4

Control Engineering

Q.1 The most basic directly achievable objective of automatic controls is -

(a) optimal performance,
(b) improve productivity,
(c) relief from repetitive manual operations,
(d) none of the above.

Q.2 Servomechanism in control systems means -

(a) feedback based control, (b) automatic control,
(c) position control system, (d) none of the above.

Q.3 Control system engineering is applicable to (i) engineering systems, (ii) biological systems, (iii) socio-economic systems.

(a) only (i) is true, (b) only (ii) is false,
(c) only (iii) is false, (d) all are true.

Q.4 A control system always has (i) a controlled variable, (ii) a manipulated variable.

(a) only (i) is true, (b) only (ii) is true,
(c) both are true, (d) both are false.

Q.5 In control systems, the word 'control' means (i) setting a desired parameter for controlled variable, (ii) measuring the value of controlled variable, (ii) applying the manipulated variable to correct or limit the deviation.

(a) all are true, (b) only (ii) is false,
(c) only (i) is false, (d) only (iii) is true.

Q.6 In control engineering, a 'plant' means (i) a complete complex of machinery, (ii) a complete machine, (iii) part of a system which can perform a particular task.

(a) all the above are true, (b) only last two are true,

(c) only last one is true, (d) none is true.

Q.7 In control engineering, a 'process' means (i) progressively continuing, (ii) series of actions or changes, (iii) leading to a particular result.

(a) only (i) is false, (b) only (i) and (ii) are true,
(c) only (i) and (iii) are true, (d) all are true.

Q.8 In conrol engineering, a 'system' means (i) a combination of components working together, (ii) performer of a certain objective,

(a) only (i) is true, (b) only (ii)is true,
(c) both are true, (d) none is true.

Q.9 In control engineering, a 'disturbance' means -

(a) a signal affecting the output,
(b) an external entity causing a change,
(c) a signal having adverse effect on output,
(d) none of theabove.

Q.10 In control engineering, 'feedback control' means -

(a) limiting the effect of disturbance by an operation on the basis of difference in input and output references,
(b) feeding back the output,
(c) comparison of input and output,
(d) none of the above.

Q.11 In control engineering, 'feedback control system' means (i) system maintaining a prescribed relationship between input and output references, (ii) comparator of input and output references, (iii) controller working on the basis of difference between input and output references.

(a) only (i) is true, (b) only (i) and (iii) are true,
(c) only (ii) and (iii) are true, (d) all are true,

Q.12 A closed loop control system, means -

(a) a feedback control system, (b) a cyclic action,
(c) a reiteration, (d) none of the above.

Q.13 An adaptive control system is which -

(a) is adapted by others,
(b) adapts itself to disturbances,
(c) is insensitive to disturbances,
(d) none of the above.

Q.14 A learning control system is that -

(a) is used for learning purposes,
(b) is able to learn from experience,

(c) conforms to both of the above,
(d) none ofthe above.

Q.15 In control engineering, transfer function concept is applicable to -
(a) linear systems only,
(b) time-invariant systems only,
(c) systems which are linear and time-invariant,
(d) systems which are linear or time-invariant.

Q.16 Transfer function of a control system is -
(a) ratio of Laplace transforms of output and input,
(b) product of response and driving function,
(c) output/input,
(d) none of the above.

Q.17 Transfer function of a control system is -
(a) a mathematical model, (b) a logical model,
(c) a physical model, (d) none of the above.

Q.18 Transfer function of a control system is -
(a) dependent upon input,
(b) dependent upon output,
(c) dependent upon input and output,
(d) independent of input and output,

Q.19 Transfer functions of various different systems -
(a) will be different, (b) can be identical,
(c) will be identical,

Q.20 Transfer function of a control system is a property of -
(a) the system,
(b) the system and the process, it carries out,
(c) input parameters,
(d) none of the above,

Q.21 Laplace transform is a technique of converting differential equations into -
(a) algebraic equations, (b) graphical representations,
(c) both of above, (d) none of above.

Q.22 A block diagram of a control system is (i) pictorial representation, (ii) functional representation, (iii) componental representation, (iv) signal flow representation.
(a) every one of above is true,
(b) all of above are true,
(c) all of above, jointly, are true.
(d) none of above is true.

Q.23 Open-loop transfer function of a control system is the ratio of -

(a) output and input signals, (b) output and error signals,
(c) feedback and error signals, (d) none of the above.

Q.24 Feedforward transfer function of a control system is the ratio of -

(a) output and input signals, (b) output and error signals,
(c) feedback and error signals, (d) none of the above.

Q.25 Closed-loop transfer function of a control system is the ratio of -

(a) output and input signals, (b) output and error signals,
(c) feedback and error signals, (d) none of the above.

Q.26 In an alternator the output voltage is affected by the loading. In a block diagram, the alternator and the loading blocks will be shown -

(a) as separate blocks, (b) as one block,
(c) anyway (a) or (b), (d) none of above.

Q.27 A signal flow graph of a control system is the representation of -

(a) signals flowing in the system,
(b) linear algebraic equations of the system,
(c) differential equations of the system,
(d) none of the above.

Q.28 In a signal flow diagram, a node is a point representing -

(a) a variable or signal, (b) branching,
(c) a process, (d) none of the above.

Q.29 In a signal flow diagram, transmittance is -

(a) gain between two nodes, (b) node transfer function,
(c) forward flow, (d) none of the above.

Q.30 A source in a signal flow diagram represents -

(a) an independent variable, (b) origination of signals,
(c) both of above, (d) none of above.

Q.31 A sink of a signal flow diagram represents -

(a) elimination of signals, (b) the output node,
(c) both of the above, (d) none of the above.

Q.32 In control engineering, 'state' means smallest number of variables that fully define the behaviour of the system for a particular input.

(a) true, (b) partially true,
(c) false,

Q.33 In control engineering, 'state space' means a space with dimensions equal to the number of state variables.

(a) true, (b) partially true,
(c) false,

Q.34 Every control system must tend to be stable.

(a) true, (b) partially true,
(c) false, (d) nonsense.

Q.35 Every useful control system must (i) have some damping, and (ii) have fast response, and (iii) reduce errors to the lowest.

(a) true, (b) partially true,
(c) false, (d) partially false.

Q.36 Control system canbe mathematically represented by (i) stste-space model, (ii) transfer function model,

(a) both of above are true, (b) none of above is true,
(c) only (i) is true, (d) only (ii) is true.

Q.37 A transfer function model of a control system is more suitable for -

(a) single-input, single-output system,
(b) multiple input/output system,
(c) optimal control system,
(d) all types of systems.

Q.38 A state-space model of a control system is more suitable for -

(a) single-input, single-output system,
(b) multiple input/output system,
(c) optimal control system,
(d) all types of systems.

Q.39 A control system is called linear if principle of superposition is applicable,

(a) true, (b) partially true,
(c) false, (d) nonsense.

Q.40 In control engineering, principle of superposition states that the response produced by the simultaneous application of two different forcing functions is the sum of the two individual responses.

(a) true, (b) partially true,
(c) false, (d) nonsense.

Q.41 A nonlinear control system is linear for small signals.

(a) true, (b) may be true,
(c) false,

Q.42 The fundamental law governing mechanical systems is -

(a) Laplace's law, (b) Kirchoff's law,
(c) Newton's second law, (d) none of the above.

Q.43 The fundamental laws governing electrical systems are -

(a) Laplace's laws, (b) Kirchoff's laws,
(c) Newton's laws, (d) none of the above.

Q.44 In control engineering, two systems are called analogus systems if (i) they can be represented by the same set of equations, (ii) they are physically same. It is true for -

(a) (i) or (ii), (b) (i)-and-(ii),
(c) (i)-and-not-(ii), (d) none of the above.

Q.45 In mechano-electrical force-voltage analogy, viscous-friction co-efficient is analogous to -

(a) capacitance, (b) capacitive reactance,
(c) resistance, (d) conductance.

Q.46 Force is to velocity, what voltage is to -

(a) charge, (b) current,
(c) resistance, (d) none of the above.

Q.47 In control engineering, analogies are,

(a) absolute, (b) vary from system to system,
(c) both of above are true, (d) none of above is true.

Q.48 D.C. servomotors, field is (i) separately excited, (ii) series excited, (iii) permanent magnet excited.

(a) all the above are true, (b) only first two are true,
(c) only last two are true, (d) none of the above is true.

Q.49 A servodriver is -

(a) driver of servomotor,
(b) servomotor-driver combination,
(c) servomotor,
(d) none of the above.

Q.50 An a.c. servomotor is an ordinary two-phase induction motor, but with -

(a) large diameter-to length ratio,
(b) small diameter-to-length ratio,
(c) large moment of inertia,
(d) none of the above.

Q.51 In an a.c. servomotor, the control (i) phase voltage is 90° out of phase of fixed phase voltage, (ii) winding is placed 90° space apart

with fixed phase, (iii) voltage is variable in magnitude and direction.

(a) all the tree are true, (b) only (i) and (iii) are true,
(c) only (ii) and (iii) are true, (d) only (i) and (ii) are true.

Q.52 A servomotor has high starting torque and low inertia for -

(a) high speed, (b) rapid acceleration,
(c) high load capacity, (d) none of the above.

Q.53 Output of a servomotor is -

(a) shaft speed, (b) shaft torque,
(c) shaft displacement, (d) none of the above.

Q.54 A flow is designated as turbulent if its Reynolds number is -

(a) greater than 3000 - 4000, (b) less than 3000 - 4000,
(c) in between 3000 - 4000, (d) none of the above.

Q.55 A flow is designated as laminar, if its Reynolds number is -

(a) greater than 3000 - 4000, (b) less than 3000 - 4000,
(c) in between 3000 - 4000, (d) none of the above.

Q.56 Heat flow is proportional to temperature difference in (i) conduction, (ii) convection, (iii) radiation.

(a) only (i) is true, (b) only (i) and (ii) are true,
(c) all are true (d) none is true.

Q.57 In radiation heat transfer is proportional to -

(a) effective temperature difference,
(b) square of effective temperature difference,
(c) cube of effective temperature difference,
(d) fourth power of effective temperature difference.

Q.58 If t_1 and t_1 are the temperatures of emitter and receiver in a heat radiatory system, the effective temperature difference is -

(a) $t_1 - t_2$, (b) $\sqrt{(t_1^2 - t_2^2)}$,
(c) $3\sqrt{(t_1^2 - t_2^2)}$, (d) $4\sqrt{(t_1^2 - t_2^2)}$,

Q.59 For a a thermal control system, thermal resistance is -

(a) change in temperature/change in heat flow rate,
(b) change in temperature difference/change in heat flow rate,
(c) change in temperature/change in heat flow,
(d) none of the above.

Q.60 For a a thermal control system, thermal capacitance is -

(a) change in temperature/change in heat flow rate,
(b) change in temperature difference/change in heat flow rate,
(c) change in heat stored/change in temperature,
(d) none of the above.

Q.61 Semiconductor strain gauge is a probe that transforms -
(a) force into voltage, (b) force into current,
(c) motion into current, (d) speed into voltage.

Q.62 For defining motion in a control system, 'degrees of freedom' means -
(a) minimum number of independent coordinates,
(b) extent of free movement,
(c) number of directions of rotation,
(d) none of the above.

Q.63 In control engineering, 'control action' means -
(a) reduction of deviation,
(b) manner of production of control signal,
(c) comparison of input and output references,
(d) none of the above.

Q.64 In control engineering, classically a 'controller' means a device that -
(a) performs the desired operation,
(b) detects the error signal and makes it effective,
(c) produces control action,
(d) none of the above.

Q.65 In control engineering, classically, an 'actuator' means a device that -
(a) actually performs,
(b) transforms control signal into a powerful input to plant,
(c) provides output,
(d) none of the above.

Q.66 The servomotor in a control system performs as a -
(a) controller, (b) actuator,
(c) sensor, (d) none of the above.

Q.67 A sensor, also known as measuring element, in control systems is put in -
(a) forwardfeed path, (b) input path,
(c) output path, (d) feedback path.

Q.68 In control systems, actual control actions are performed by -
(a) sensors, (b) actuators,
(c) controllers, (d) any of the above.

Q.69 In an on-off or two-position control action, the actuator has -
(a) two or three positions, (b) two positions,
(c) continuity, (d) any of the above.

Q.70 In a two-position control system, the frequency of changeovers between the two positions, is increased by -

(a) increasing the differential gap,
(b) decreasing the differential gap,
(c) by any of the above,
(d) by none of the above.

Q.71 In a proportional control system, the proportional gain constant is the ratio of -

(a) input and output,
(b) error signal and plant output,
(c) output of the controller and actuating error signal,
(d) actuator output and input.

Q.72 In control systems, proportional control action is performed by -

(a) actuator,
(b) controller amplifier with variable gain,
(c) feedback sensor,
(d) none of the above.

Q.73 In integral control systems -

(a) controller output is proportional to rate of change of error,
(b) rate of change of controller output is proportional to error,
(c) any one of the above,
(d) none of the above.

Q.74 In control engineering, in a derivative control action -

(a) controller output is proportional to rate of change of error,
(b) rate of change of controller output is proportional to error,
(c) any one of the above,
(d) none of the above.

Q.75 In control systems, as compared to pneumatic operating pressure, hydraulic operating pressure is -

(a) much higher, (b) slightly higher,
(c) equal, (d) lower.

Q.76 In control systems, as compared to pneumatic operating power, hydraulic operating power is -

(a) much higher, (b) slightly higher,
(c) equal, (d) lower.

Q.77 In a proportional control system, a steady state error, called offset, occurs in response to a step input. This error can be eliminated by introducing in the system, -

(a) a derivative control action, (b) an integral control action,

(c) any one of the above, (d) none of the above.

Q.78 In control systems, with the use of a derivative control actions, system stability and accuracy -

(a) improve, (b) deteriorate,
(c) remain uneffected,

Q.79 In a control system, a derivative control action can be used -

(a) alone,
(b) only in combination with other types of actions,

Q.80 In control systems, feedback (i) reduces sensitivity to parameter variation, (ii) eliminates unwanted disturbances.

(a) both are true, (b) only (i) is true,
(c) only (ii) is true, (d) none is true.

Q.81 In a closed loop control systems, a system may become unstable with the use of a -

(a) negative feedback, (b) positive feedback,
(c) any one of above, (d) none of above.

Q.82 To test control systems, test signals of the type (i) step, (ii) ramp, (iii) acceleration, (iv) impulse, (v) sinusoidal, functions are used -

(a) all the above are true, (b) only first two are true,
(c) only last two are true, (d) none is true.

Q.83 Selection of test signal depends upon -

(a) nature of the transfer function,
(b) nature of feedback,
(c) nature of expected input to system,
(d) all of the above.

Q.84 For a control system, if the input is expected to be gradually varying with time, the most suitable test signal for the system will be -

(a) step function, (b) ramp function,
(c) impulse function, (d) none of the above.

Q.85 For a control system, if the input is expected to be of suddenly varying nature, the most suitable test signal for the system will be -

(a) step function, (b) ramp function,
(c) impulse function, (d) none of the above.

Q.86 For a control system, if the input is expected to be of a shocking nature, the most suitable test signal forthe system will be -

(a) step function, (b) ramp function,
(c) impulse function, (d) none of the above.

Q.87 A control system is called to be in equilibrium if, in the absence of an input or disturbance -
(a) the output stays in the same state,
(b) the system appears dead,
(c) the output is zero,
(d) none of the above.

Q.88 A linear time-invariant control system is stable if, on being subjected to a disturbance, the output comes back to -
(a) zero, (b) equilibrium,
(c) any of the above, (d) none of the above.

Q.89 If output of a system at steady state does not exactly agree with the input, the system is said to have (i) steady-state error, (ii) inaccuracy, (iii) failed.
(a) only (i) is true, (b) only (i) and (ii) are true,
(c) only (iii) is true, (d) only (ii) is true.

Q.90 Causes of steady-state error of a control system are (i) change in nature of reference input, (ii) system component imperfections.
(a) only (i) is true, (b) only (ii) is true,
(c) both are true, (d) none is true.

Q.91 Impulse-response function of a control system is output of the system to -
(a) an impulse input,
(b) a unit impulse input when initial conditions are zero,
(c) a unit step input,
(d) none of the above.

Q.92 Impulse-response function of a control system is (i) known as weighting function, (ii) provides information on dynamic characteristics, (iii) same as that due to a very short-duration pulse input.
(a) all the above are true, (b) only first two are true,
(c) only (ii) is true, (d) only (iii) is true.

Q.93 In control engineering, an impulse function and a pulse function of large amplitude and narrow width, are -
(a) same, (b) approximately same,
(c) different,

Q.94 Response of all systems having the same transfer function and with the same input function, will be -
(a) same,
(b) dependent upon the nature of the system,

(c) dependent upon the nature of the input,
(d) none of the above.

Q.95 The output of a first order system, in response to a unit-step function input, at a time equal to the time constant, will be -
(a) 1, (b) 1.63,
(c) 0.63, (d) none of the above.

Q.96 If, in a first order control system, for an input of unit-step function, the initial rate of rise of response is maintained, the output will reach its final value in a time interval of -
(a) nearly 4 times the time-constant,
(b) time constant,
(c) half the time constant,
(d) none of the above.

Q.97 As a test for a control system to be of first order or not, the trace of log|c(t) -c(_)| againt time t, where c(t) is the time response of the system for an unit-step function, will be -
(a) exponential curve, (b) straight line,
(c) circular, (d) none of the above.

Q.98 The steady-state unit-ramp response of a first order control system, has an error in following the input, equal to -
(a) transfer function, (b) time-constant,
(c) square of time-constant, (d) none of the above.

Q.99 For a first order control system (i) response to the derivative of a signal is the derivative of the response of the original signal, (ii) response to the integral of a signal is the integral of the response to the original signal.
(a) only (i) is true, (b) only (ii) is true,
(c) both are true, (d) none is true.

Q.100 A second order control system, the fastest response can be obtained by keeping the damping ratio between -
(a) 0.1 and 0.4, (b) 0.5 and 0.8,
(c) 1.0 and 2.0, (d) none of the above.

Q.101 Critical damping of a control system means the point where oscillations in the output of the system -
(a) have just become zero, (b) are zero,
(c) are uniform, (d) none of the above.

Q.102 The delay time in the transient response of a control system to a unit-step input, is the time taken by te response to reach -
(a) 0.33 times the final value, (b) 0.5 times the final value,
(c) 0.63 times the final value, (d) the final value.

Q.103 The peak-time in the transient response of a control system to a unit step input, is the time taken by the response to reach -

(a) the final value, (b) the first peak value,
(c) steady state. (d) none of the above.

Q.104 The transient response of an underdamped second order control system is -

(a) zero, (b) oscillatory about zero,
(c) unit-step, (d) none of the above.

Q.105 In a second order control system, the velocity feedback give the same improvement in response as available by proportional-plus-derivative control action.

(a) true, (b) false,
(c) nonsense.

Q.106 For a control system to be absolutely stable, according to Routh's criterion, all the coefficients of the denominator polynomial of transfer function, must be -

(a) nonzero, (b) positive,
(c) nonzero and positive, (d) none of the above.

Q.107 For a type zero control system, the steady-state error to a unit-step response -

(a) can not be zero, (b) will be zero,
(c) may be zero (d) none of the above.

Q.108 For a control system, as the type number increases, its stability problems -

(a) increase, (b) decrease,
(c) are unaffected,

Q.109 The type number of a control system, indicates the number of -

(a) system components,
(b) integrations in the feedforward path,
(c) integrations and differentiations,
(d) none of the above.

Q.110 Static position error constant for a control system is finite if the system is of type -

(a) zero, (b) finite,
(c) infinite, (d) none of the above.

Q.111 If a control system is of type higher than zero, its static position error constant will be -

(a) zero, (b) finite,
(c) infinite, (d) none of the above.

Q.112 For all control systems, except those of zero type, the steady-state error for a step input is -

(a) zero, (b) finite,
(c) infinite, (d) none of the above.

Q.113 The static velocity error constant for a control system is -

(a) error in position due to ramp input,
(b) error in velocity of motion,
(c) error in rate of change,
(d) none of the above.

Q.114 The steady-state actuating error for unit-ramp input of a control system is (i) infinite for zero type system, (ii) zero for type 2 or higher system.

(a) both are true, (b) both are wrong,
(c) only (i) is true, (d) only (ii) is true.

Q.115 In a type 1 control system, a ramp input response is the same velocity and -

(a) zero errorin position, (b) finite error in position,
(c) infinite error in position, (d) none of the above.

Q.116 Static acceleration error constant for a type zero and one control system (ie parabolic input response) is -

(a) zero, (b) finite,
(c) infinite, (d) none of the above.

Q.117 Static acceleration error constant for a type two control system (ie parabolic input response) is -

(a) zero, (b) finite,
(c) infinite, (d) none of the above.

Q.118 Static acceleration error constant for a type three or higher control system (ie parabolic input response) is -

(a) zero, (b) finite,
(c) infinite, (d) none of the above.

Q.119 The performance index of a control system is a number that indicates -

(a) efficiency of the system,
(b) merits of the system,
(c) goodness of the system performance,
(d) none of the above.

Q.120 An optimal control system will have its performance index (i) minimum, (ii) maximum.

(a) only (i) is true,
(b) only (ii) is true,

(c) both are true depending upon the situation,
(d) none of the above.

Q.121 The performance index of a control system must be a function of -
(a) the parameters of the system,
(b) system response,
(c) system efficiency,
(d) none of the above.

Q.122 For an integral square-error(ISE) criterion of performance, optimization leads to -
(a) initial errors are weighed heavily,
(b) late transient errors are weighted heavily,
(c) both of above,
(d) none of the above.

Q.123 For an integral of time-multiplied square-error (ITSE) criterion of performance, optimization leads to -
(a) initial errors are weighted heavily,
(b) late transient errors are weighted heavily,
(c) both of above,
(d) none of the above.

Q.124 For an integral absolute-error (IAE) criterion of performaonce optimization -
(a) fuel efficiency is the best case,
(b) power consumption can be minimized,
(c) both of above can be minimized,
(d) none of the above.

Q.125 In a control system closed-loop poles are -
(a) values of specified parameters,
(b) roots of characteristic equation,
(c) type number of the system,
(d) none of the above.

Q.126 The basic characteristic of the transient response of a control system is closely related to the closed-loop poles. This concept is exploited in -
(a) Runge-Kutta method, (b) Root-Locus method,
(c) both of above, (d) none of the above.

Q.127 The characteristic equation of a control system is obtained by equating to zero, the -
(a) numerator of the transfer function,
(b) denominator of the transfer function,

(c) transfer function,
(d) none of the above.

Q.128 In Root-Locus method, closed-loop roots of the characteristic equation of a control system, are those values of s which satisfy -

(a) only angle condition,
(b) only magnitude condition,
(c) both angle and magnitude conditions,
(d) none of the above.

Q.129 The root-loci of a control system, as used in Root-Locus method, are symmetrical about -

(a) real axis, (b) imaginary axis,
(c) both of above, (d) none of the above.

Q.130 For a control system, if G(s) is the feedforward path gain and H(s), the feedback path gain, the closed-loop transfer function is -

(a) G(s)/[1 + G(s)H(s)], (b) G(s)/H(s),
(c) G(s)H(s), (d) none of the above.

Q.131 For a control system, if G(s) is the feedforward path gain and H(s), the feedback path gain, the open-loop transfer function is -

(a) G(s)/[1 + G(s)H(s)], (b) G(s)/H(s),
(c) G(s)H(s), (d) none of the above.

Q.132 For a control system, if G(s) is the feedforward path gain and H(s), the feedback path gain, the characteristic equation is -

(a) G(s)/[1 + G(s)H(s)] = 0, (b) 1 + G(s)/H(s) = 0,
(c) 1 + G(s)H(s) = 0, (d) none of the above.

Q.133 A control system is stable, if its closed-loop poles are located on the s-plane on -

(a) right side imaginary axis, (b) left side of imaginary axis,
(c) upper side of real axis, (d) lower side of real axis.

Q.134 A second order system is -

(a) always stable, (b) can be unstable,
(c) can be stable or unstable,

Q.135 A minimum phase control system is that which has all the poles and zeros in the -

(a) left-half s-plane, (b) right-half s-plane,
(c) upper-half of s-plane, (d) lower-half of s-plane.

Q.136 A nonminimum phase control system is that which has one or more poles or zeros in the -

(a) left-half s-plane, (b) right-half s-plane,
(c) upper-half of s-plane, (d) lower-half of s-plane.

Q.137 A transport-lag or dead-time of a control system is (i) delay in measurement, (ii) delay in controller action, (iii) delay in actuator operation, (iv) delay in plant effect.

(a) only (i) is true, (b) only first two are true,
(c) only first three are true, (d) all are true..

Q.138 Practical process control systems have dead time.

(a) mostly true, (b) mostly false,
(c) can be true or false equally.

Q.139 A dead-time in a control system can produce system instability in -

(a) all systems,
(b) all but first order systems,
(c) all above second order systems,
(d) none of the above.

Q.140 Root contours of a control system, are root locii obtained with variation of -

(a) two or more parameters, (b) gain,
(c) transport lag, (d) none of the above.

Q.141 By frequency response of a control system, we mean -

(a) steady-state response to sinusoidal input,
(b) transient response to sinusoidal input,
(c) both (a) and (b),
(d) none of the above.

Q.142 Frequency response and transient response of a control system are corelated only indirectly, for -

(a) all types of systems,
(b) all types, except zero-order systems,
(c) all types, except first order systems,
(d) all types, except second order systems.

Q.143 The output of a linear time-invariant control system, to a sinusoidal input signal will be of -

(a) same frequency as that of input,
(b) half frequency as that of input,
(c) double frequency as that of input,
(d) none of the above.

Q.144 The sinusoidal transfer function of a linear system is obtained from the transfer function of the system by substituting jw, where w is the frequency, for -

(a) s, (b) s^2,
(c) forward transfer function, (d) open-loop transfer function.

Q.145 A Bode diagram of a control system, consists two curves, one is phase angle of sinusoidal transfer function, against frequency on log scale, the other is also for the sinusoidal transfer function, but for -

(a) magnitude against frequency,
(b) log of magnitude against frequency,
(c) any of the above (a) or (b),
(d) none of the above.

Q.146 The polar plot of sinusoidal transfer function of a control system is a plot on polar coordinates, with variation in frequency from zero to infinity, of magnitude versus -

(a) frequency, (b) gain,
(c) phase angle, (d) none of the above.

Q.147 For two control systems connected in cascade, the overall transfer function is -

(a) the sum of the two transfer functions,
(b) the product of the two transfer functions,
(c) under-root of the sum of the squares of the two transfer functions.
(d) none of the above.

Q.148 The Nichols plot of sinusoidal transfer function of a control system is a plot on polar coordinates, with variation in frequency from zero to infinity, of log of magnitude versus -

(a) frequency,
(b) gain,
(c) phase angle or phase margin,
(d) none of the above.

Q.149 Nyquist path is the closed contour in the s-plane covering the entire right half of the plane.

(a) true, (b) partially true,
(c) false, (d) nonsense.

Q.150 If open-loop transfer function G(s)H(s) has k poles in the right-half s-plane, then, according to Nyquist stability criterion, for stability, the G(jw)H(jw) locus, as w varies from $-\infty$ to $+\infty$, must encircle (-1 + j0) point -

(a) k^2 times in clockwise direction,

(b) k times in clockwise direction,
(c) k times in counter-clockwise direction,
(d) none of the above.

Q.151 What modification is necessary in the Nyquist stability criterion, if the poles or zeros of G(s)H(s) lie on vertical axis.
(a) Nothing,
(b) Nyquist path excludes the origin,
(c) Nyquist path excludes the whole vertical axis,
(d) none of the above.

Q.152 For linear control systems using Nyquist stability criterion, if there is no encirclement of (-1 +j0) point and there are no poles of G(s)H(s) in the right-half s-plane,
(a) the system is unstable,
(b) the system is stable,
(c) nothing can be said on this basis.

Q.153 For linear control systems using Nyquist stability criterion, if the counterclockwise encirclements of (-1 +j0) point are equal to the poles of G(s)H(s) in the right-half s-plane,
(a) the system is unstable,
(b) the system is stable,
(c) nothing can be said on this basis.

Q.154 For linear control systems using Nyquist stability criterion, if encirclements of (-1 +j0) point are clockwise,
(a) the system is unstable,
(b) the system is stable,
(c) nothing can be said on this basis.

Q.155 Nyquist stability criterion is applicable to inverse polar plot also with substitution of 1/G(s)H(s) for -
(a) G(s)H(s), (b) 1 + G(s)H(s),
(c) none of the above.

Q.156 For investigating relative stability of linear control systems by Nyquist stability criterion, the Nyquist path -
(a) needs no modification, (b) needs to be modified,

Q.157 In control engineering, compensation means -
(a) to provide the system additional power for its losses,
(b) to modify system dynamics to satisfy the given specification,
(c) both of the above,
(d) none of the above.

Q.158 The performance specifications of a control system are related to (i) accuracy, (ii) relative stability, (iii) speed of response.
(a) all of the above are true, (b) only first two are true,
(c) only last two are true, (d) none of the above.

Q.159 In control engineering, a compensator means a dvice inserted into the system -
(a) for providing power lost in the system,
(b) for bypassing loss-making components,
(c) satisfying the given specifications,
(d) none of the above.

Q.160 In control systems, a series compensator is placed -
(a) in series with the feedback transfer function,
(b) in series with the feed forward transfer function,
(c) in series with the input,
(d) in series with the output.

Q.161 In control systems, a parallel compensation or feedback compensation, is placed -
(a) in the feedback path of the system,
(b) to provide additional feedback path in the system,
(c) any of the above,
(d) none of the above is true.

Q.162 In compensating a control system, additional amplifier may be necessary, if -
(a) series compensation is chosen,
(b) parallel compensation is chosen,
(c) any of the above is chosen,

Q.163 In control engineering, a lag network is that which provides phase-lag in the output wrt. the input, in the response of -
(a) a pulse input, (b) a sinusoidal input,
(c) an impulse input, (d) none of the above.

Q.164 In control engineering, a lag-lead network provides the output in response to a sinusoidal input, which -
(a) lags the input but has a leading power factor,
(b) has a lagging power factor but leads the input,
(c) lags or leads the input at different frequencies,
(d) none of the above is true.

Q.165 In control systems, addition of an integral control adds a -
(a) pole at the origin, (b) zero,
(c) any of the above, (d) none of the above.

Q.166 By addition of a pole to the open-loop transfer function of a control system, its relative stability is -

(a) improved, (b) impaired,
(c) unaffected.

Q.167 Addition of a derivative control in the feedforward path of a control system, adds a zero in the -

(a) closed-loop transfer function,
(b) open-loop transfer function,
(c) in both of above,
(d) in none of above.

Q.168 By addition of a zero in the open-loop transfer function of a control system,

(a) its stability and transient response are improved,
(b) its stability and transient response are impaired,
(c) its stability improves but transient response is impaired,
(d) its stability is impaired but transient response is improved.

Q.169 In control systems phase margin, gain margin and resonant peak magnitude provide an indication of -

(a) system performance, (b) system damping,
(c) system transient response, (d) non of the above.

Q.170 In control systems gain crossover frequency, resonant frequency and bandwidth provide an indication of -

(a) system performance, (b) system damping,
(c) system transient response, (d) non of the above.

Q.171 The information on closed-loop behaviour of a control system, provided by the open-loop frequency response of the system are (i) steady-state behaviour from the low-frequency region, (ii) relative stability from the medium-frequency region, (iii) complexity from the high-frequency region.

(a) all the above are true, (b) only (i) and (ii) are true,
(c) only (ii) and (iii) are true, (d) none of the above is true,

Q.172 For having a satisfactory relative stability from a control system, the open-loop frequency response (i) in low frequency range should have a high gain, (ii) in high frequency range, the attenuation should be fast.

(a) all the above are true, (b) only (i) is false,
(c) only (ii) is true, (d) none of the above is true,

Q.173 In control systems, an R-C network and a mechanical spring-dashpot network can be used as -

(a) lag compensators. (b) lag or lead compensator,

(c) lead compensators, (d) none of the above.

Q.174 In control systems, an operational amplifier (opamp) can be used as a -

(a) lag compensator, (b) lead compensator,
(c) lag or lead compensator, (d) none of the above.

Q.175 In control systems, a lag compensator basically provides -

(a) phase lag characteristic,
(b) high-frequency attenuation characteristic,
(c) low-frequency attenuation characteristic,
(d) none of the above.

Q.176 In control systems, a lead compensator (i) increases bandwidth, (ii) speeds up response, (iii) decreases the maximum overshoot in step response.

(a) all the above are true, (b) only first two are true,
(c) only last two are true, (d) none of the above is true.

Q.177 In control systems, a lag compensator basically (i) increases low-frequency gain, (ii) reduces speed of response.

(a) all the above are true, (b) only first two are true,
(c) only last two are true, (d) none of the above is true.

Q.178 In control systems, usually a lag-lead compensator adds -

(a) two zeros only, (b) two poles and two zeros,
(c) one pole and one zero, (d) two poles only.

Q.179 In control systems, an R-C parallel network connected in series, works as a -

(a) lag compensator, (b) lead compensator,
(c) lag-lead compensator, (d) none of the above.

Q.180 In control systems, an R-C network connected in shunt in the path, works as a -

(a) lag compensator, (b) lead compensator,
(c) lag-lead compensator, (d) none of the above.

Q.181 In control systems, a PD controller is a -

(a) lag controller, (b) lead controller,
(c) lag-lead controller, (d) none of the above.

Q.182 In control systems, a PI controller is a -

(a) lag controller, (b) lead controller,
(c) lag-lead controllor, (d) none of the above.

Q.183 In control systems, a PID controller is a -

(a) lag controller, (b) lead controller,
(c) lag-lead controller, (d) none of the above.

Q.184 In control engineering, Ziegler-Nichols rules are applicable to -
(a) get transfer function characteristics,
(b) improve stabilty,
(c) tune PID controllers,
(d) none of the above.

Q.185 In control systems, an I-PD controller is a controller -
(a) same as PID controller,
(b) in which proportional and derivative controls are put in feedback path,
(c) in which integral control made dominant,
(d) none of the above.

Q.186 In nonlinear control systems, as the amplitude of oscillations increases, the frequency of oscillations -
(a) increases, (b) decreases,
(c) any of the above, (d) none of the above.

Q.187 For a nonlinear control system, the complex ratio of fundamental harmonic component of output to sinusoidal input, is known as -
(a) complex transfer function, (b) ratio function,
(c) describing function, (d) none of the above.

Q.188 In a nonlinear control system, if the output is a function of input amplitude with a phase-lag, the system nonlinearity is called -
(a) on-off nonlinearity with hysterisis,
(b) amplitude-dependent nonlinearity,
(c) phase-derived nonlinearity,
(d) none of the above.

Q.189 In a nonlinear control system, if the output is zero for certain levels of input, it is known as -
(a) sensitivty-base nonlinearity,
(b) dead-zone or threshold nonlinearity,
(c) zero-factor nonlinearity,
(d) none of the above.

Q.190 In a nonlinear control system, if the output is flattened for higher amplitude of input, it is known as -
(a) top-flat nonlinearity, (b) saturation nonlinearity,
(c) limitation nonlinearity, (d) none of the above.

Q.191 For two nonlinear control elements connected in series, the transfer function of the adjoint can be obtatained by multiplying the two individual transfer functions, provided -
(a) two elements are identical,

(b) inputs to the two elements are sinusoidal,
(c) inputs to the two elements are linear,
(d) none of the above.

Q.192 A periodic wave with slight distortion from sinusoidality can be made sinusoidal by passing it through a -
(a) high-pass filter, (b) low-pass filter,
(c) attenuator, (d) amplifier.

Q.193 A feedback control system, that brings nonzero states, caused by external disturbances, to the origin with good speed, is called -
(a) feedback control system, (b) regulator system,
(c) zero-pull system, (d) none of the above.

Q.194 By differentiation in a control system, signal-to-noise ratio is -
(a) increased, (b) decreased,
(c) unchanged,

Q.195 In control engineering, state-observation means -
(a) estimation of unmeasurable state variables,
(b) observation of system behaviour,
(c) feedback,
(d) none of the above.

Q.196 A control system with one integrator in the feedforward path is called -
(a) type I system, (b) first order system,
(c) zero order system, (d) none of the above.

Q.197 Type of a control system is determined by the number of control elements, in the feedforward path, that cause -
(a) integration, (b) differentiation,
(c) proportionalization, (d) none of the above.

Q.198 In a model-reference control system, the outputs of the real system and that of an idealized model (mathematical or computerized) are compared to generate the -
(a) control signal, (b) error signal,
(c) input signal, (d) none of the above.

Q.199 An adaptive control system, to maintain optimal performance regardless of the environmental changes, measures its own performance and modifies it if necessary by using as control signal, the deviation between its own performance and the -
(a) given idealized performance, (b) past performance,
(c) none of the above.

Q.200 The performance index of a control system, is desired to have the qualitie of reliability, selectivity and applicability.

(a) true,
(b) partially true,
(c) incomplete,
(d) none of the above.

Q.201 The highest quality of system so far conceived is -

(a) adaptive loop system,
(b) feedback loop system,
(c) learning loop system,
(d) none of the above.

ABSTRACT

4.1. BASIC DEFINITIONS

Controlled variable : The quantity or condition that is measured and controlled by a control system.

Manipulated variable : The quantity or condition that is varied by the controlled so as to affect the value of the controlled variable.

Control : Measuring the value of the controlled variable of the system and applying the manipulated variable to the system to correct or limit the deviation of the measured value of the controlled variable from its desired value.

Plant : A set of machinery functioning together with a purpose to perform a particular operation.

Process : A natural or an artificial or voluntary, progressively continuing operation or development marked by a series of gradual changes that succeed one another in a relatively fixed way and lead toward a particular result or end.

System : A combination of components that act together and perform a certain objective. It may be a physical entity or an abstract phenomenon.

Disturbance : A signal that tends to adversely affect the value of the output of a system. It may be internal or external to the system.

Feedback control : An operation that tends to reduce difference between the output of a system and some reference input, in the presence of a disturbance.

Feedback control system : A system that maintains a prescribed relationship between the output and some reference input by comparing them and using the difference as a means of control.

Servo system : A feedback control system in which the output is some mechanical position, velocity or acceleration.

Automatic regulating system : A feedback control system in which the reference input or the desired output is either constant or slowly varying with time and in which the primary task is to maintain the actual output at the desired value in the presence of disturbances.

Process control systems : An automatic regulating system in which the output is a variable such as temperature, pressure, flow etc. (a process variable).

Closed-loop control system : Feedback control system.

Open-loop control system : A system in which the output has no effect on the control action.

Adaptive control system : A system which is capable of self-adjusting or self-modifying in accordance with unpredictable changes in conditions of environment or structure.

Learning control system : A system which is capable of learning by experiences.

Linear control system : A system to which, the principle of superposition applies.

Nonlinear control system : A system to which the principle of superposition does not apply.

Time-invariant control system : A system whose parameter do not change with time. In such systems, response to an input does not depend upon the time of applying the input.

Time-variant control system : A system whose one or more parameters change with time. In such systems, response to an input may depend upon the time of applying the input.

Continuous-time control system : A system in which system variables are a function of continuous time.

Discrete-time control system : A system whose one or more variables are known only at discrete instants of time.

Lumped-parameter control system : System that can be described by ordinary differential equations.

Distributed parameter control system : A system that need partial differential equations to describe it.

Deterministic control system : A system whose response to an input is predictable and repeatable.

Stochastic control system : A system whose response to an input is not predictable and repeatable.

Causal system : A system is said to be causal if and only if the output at any time t, Y(t) depends only on the input at times upto and including time t. A real system must be causal.

State space approach : A description of a system in which the state gives a complete description of the system at a given time. It implies that there are orderly rules for the transition from one state to another.

4.2. CONTROL SYSTEM

A control system may be represented by Fig. 4.1(a). If the systems output reference signal is not feedback to the input, the system is said to be an open loop system. If the output response signal is feedback to the input side, the system is called a closed-loop control system. This is shown in Fig. 4.1(b).

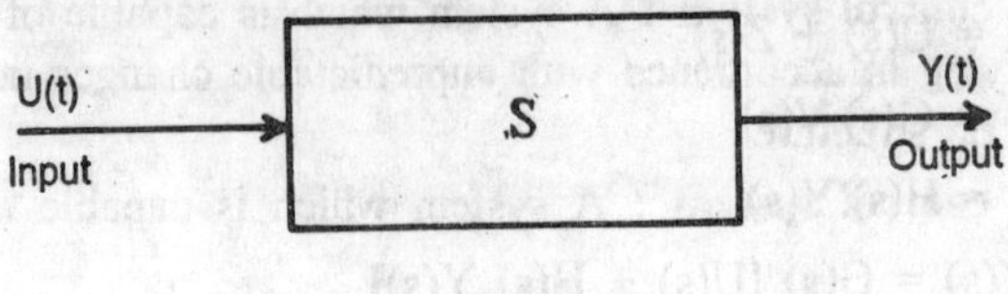

(a)

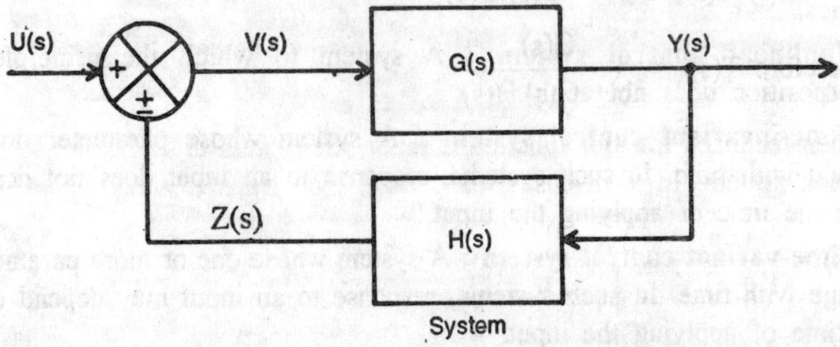

(b)

Fig. 4.1. (a) Control system representation. (b) Closed-loop control system.

Transfer function

Transfer function of a linear time invariant (LTI) system is defined as the Laplace transform of systems response to a unit impulse. Let h(t) be the response such that h(t) = 0 for t < 0. The Laplace transform for h(t) is given by :

$$H(s) = \int_{\infty}^{\infty} h(t)\, e^{-st}\, dt$$

H(s) is known as the transfer junction of the system. For any input u(t), whose Laplace transform is u(s), the Laplace transform of the output is given by:

$Y(s) = H(s)\, U(s)$

For a close-loop system, shown in Fig. 4.1(b) with signal values as shown. G(s) is known as the forward transfer function and is given by :

$G(s) = Y(s)/V(s)$

H(s) is the feedback transfer function give by :

$H(s) = Z(s)/Y(s)$

The product G(s).H(s) is referred to as the open-loop transfer function of the system.

The closed-loop transfer function of this system is :

$$T(s) = \frac{Y(s)}{U(s)}$$

Since $V(s) = U(s) \pm Z(s)$

$Y(s) = G(s).V(s)$

$Z(s) = H(s).Y(s)$

We have $Y(s) = G(s) [U(s) \pm H(s) Y(s)]$

or $[1 + G(s).H(s)] Y(s) = G(s).U(s)$

$$\text{Therefore } T(s) = \frac{G(s)}{1 \pm G(s).H(s)}$$

When feedback transfer function is unity, the system is said to be a unity feedback system. In such cases, the closed loop transfer function is :

$$T(s) = \frac{G(s)}{1 \pm G(s)}$$

Characteristic polynomial

A generalized differential equation form representing a control system is :

$$\frac{d^nY(t)}{dt^n} + a_{n-1}\frac{d^{n-1}Y(t)}{dt^{n-1}} + \ldots + a_o Y(t)$$

$$= b_m \frac{d^nU(t)}{dt^m} + \ldots + b_o u(t)$$

where U(t) and Y(t) are the input and output respectively, and a_i, i = 0, 1, 2, ... n–1, and b_j, j = 0, 1, 2, . . ., m are real numbers.

Taking Laplace transforms of both sides of the above equation, we get :

$$(S^n + a_{n-1}S^{n-1} + \ldots + a_o) Y(s) = (b_mS^m + \ldots + b_o) U(s)$$

$$\text{or } Y(s) = \frac{b_mS^m + b_{m-1}S^{m-1} + \ldots + b_o}{S^n + a_{n-1}S^{n-1} + \ldots + a_o} U(s)$$

$$\text{or } T(s) = \frac{Y(s)}{U(s)} = \frac{b_mS^m + b_{m-1}S^{m-1} + \ldots + b_o}{S^n + a_{n-1}S^{n-1} + \ldots + a_o}$$

The denominator of the transfer function is known as the characteristic polynomial. The roots of the characteristic polynomial are known as the poles of the transfer function. Roots of the numerator polynomial of the transfer function are known as the zeros of the transfer function.

4.3. CONTROL SYSTEM COMPONENTS

A closed loop control system is shown in Fig. 4.2 representing main components of a control system and their interconnections.

(a) Error detector or summer

The first component of a control system is the error detector which compares the control reference signal, obtained from the system operator, and the feedback signal and if there exists some appreciable difference between the two, an actuating signal is provided to the controller. The control reference signal may come from another control system in command of the present system or from a system operator. Obviously the feedback signal and the control reference signal have same dimensions. In some cases, the error detector is in-built in the controller.

(b) Controller

The actuating signal received from the error detector is usually a weak signal. Hence it becomes necessary for the controller to amplify this signal to an appreciable value. Therefore controllers usually have an amplifier, as the first stage. The amplified signal is then transformed to a manipulated variable, which, in most of the cases, is different from the control action signal. The manipulated variable may be a signal proportional to the control action signal or it may be a digital signal to put the actuator ON or OFF depending upon the control action signal. For example, in an automatic temperature control system, heaters are switched ON/OFF through a power contactor. In this case the operating coil supply voltage of the contractor is the manipulated variable, which is present or absent depending upon the actuating signal.

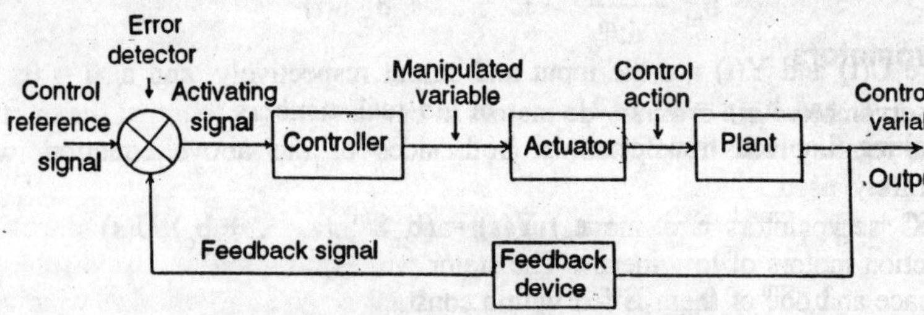

Fig. 4.2. Basic control system.

(c) Actuator

The manipulated variable is fed to the actuator which is a buffer device between the controller and the plant. It may be a contactor, solenoid valve, servomotor or such, usually, electromechanical device. It converts the manipulated variable to a powerful action known as the control action. Therefore an actuator is a device to amplify and convert the form of energy.

(d) Plant

It may be argued that in most cases the plant is not a part of the control system but the complex engineering system on which the control action is to be applied.

Apart from the control action as input of the plant, plant provides the input to the feedback device in the form of the controlled variable. Hence the plant, usually, physically carries the feedback device to provide it the necessary input. In a speed control system, the tachogenerator is mounted on the speeding shaft, in the case of a temperature control system, the thermometer is mounted nearest to the hot spot.

(e) Feedback device

When the plant has been controlled, it becomes necessary to know whether the controlled variable is within the specified limits or some more correcting action is required. Continuous monitoring and control of the controlled variable also make it necessary that the feedback element keeps a watch on it and acts accordingly.

The feedback device converts the controlled variable to the form comparable to the control reference signal, which is in most of the cases, a voltage signal, since we are referring to electrical control systems. A technogenerator, a limit switch, a pyrometer, a pressure switch etc. are the feedback devices.

4.5. CONTROL SYSTEM DEVICES

The devices used in a control system, mostly perform the functions of actuators and the feedback. For these applications, some more frequently used devices are discussed below.

Servomotors

DC servomotors are ordinary dc motors in construction except that the inertia of the rotating mass is kept minimum for a faster response. DC servomotors are rarely used.

AC servomotors are more rugged and robust. These are two phase induction motors of low inertia. The stator winding two phases are displaced in space and one of them is fed with a constant ac voltage. Another winding, known as the control winding is fed with a controlled supply required according to the motor rotation desired.

The fixed winding not only provides rotational power but also aids in braking when the control winding supply is zero, bringing the motor to quick stoppage.

Amplidynes

These are rotating amplifiers with gain of several thousand. The amplification of the field supply is obtained in two stages of dc generation in the armature. In the first stage, a strong cross-magnetization field by the armature current, by short-circuiting the two cross-axis carbon brushes in the commutator, is obtained. The second stage utilizes, the cross-magnetizing field as the excitation of dc voltage generation in the direct axis carbon brushes on the commutator, which is taken as the output of the amplifier.

Tachogenerators

To provide a signal proportional to the speed of a machine, tachogenerator are used. These are of ac as well as dc type but due to simplicity and robustness, ac tachogenerators are more commonly used. These are permanent magnet ac alternators providing a single phase output whose frequency and voltage is proportional to the speed. The output is usually rectified to get a proportional dc voltage for use in control systems. Techogenerator is coupled to the rotating shaft, whose speed is to be feedback to the control system.

Synchros or selsyns

Synchros provide a remote availability of angular position of a shaft, through conversion of mechanical displacement to electrical signals and back to mechanical displacement. Two types of synchrosets are in use. In the first type two identical synchos form a set. A synchro is a rotating machine with rotor having a salient pole construction with single phase ac winding. The stator has a three phase winding. The rotor is fed with a fixed ac supply and coupled to the shaft whose position is to be transmitted. This synchro is said to be the transmitter. The other synchro, known as motor, placed at a remote place, has the same ac supply for the rotor as the transmitter synchro rotor. The stator windings to the two synchros are connected together. In such a system the rotors of the two synchros will always be aligned identically with the stators.

In another synchro system, which provides the displacement equal to the sum or difference of the two shaft positions, a third motor synchro, having three phase windings on stator and rotor, is used as the controlled device. The two controlling synchros are of the first type, coupled to the two shafts.

Stepper motor

Stepper motor takes a fixed step of rotation for each pulse supplied to it. The angle of step may be a degree to half rotation. The stator has multiple phase windings and one phase winding is energized at a time by the controlling pulse. The rotor, having variable reluctance construction may or may not have a permanent magnet. On the supply of a pulse to a stator winding, the rotor tends to align in a position of minimum reluctance with the energized stator winding. These motors are used in digital controls.

Gyroscope

Gyroscopes provide the angular displacement of the machines whose bases are not fixed but float in space like the ships in the ocean and the planes in the air. Their basis of working is that a spinning mass tends to maintain its axis in a fixed direction, unless disturbed by a force.

Gyroscope has a spinning mass rotating in a horizontal or vertical direction, and suspend on a set of gimbles, the outermost fixed on the base of the machine. When the machine tilts or rotates, the spinning mass, maintaining its original axis of rotation, tilts the gimbles providing a signal indicating the machine position.

Hydraulic/pneumatic cylinders

These devices convert a flow into mechanical displacement by the displacement of a piston in a cylinder with the thrust of oil or air. With regular supply of oil/air, the displacement may be a function of pressure.

Hydraulic pump

A rotation may be converted to a pressure or volume or even a displacement by the help of a hydraulic pump, which are of various designs and contruction.

4.6. ANALOGY IN CONTROL SYSTEMS

Control systems which can be represented by the same mathematical model are known as analogous systems. This greatly helps in analysis of various types of control systems by comparison or through mathematical modelling of one of them. It is a notable fact that a system cannot have more than one mathematical equivalent model while a mathematical model can have various equivalent control systems, one electrical, one mechanical, one electronic etc.

Analogous quantities of some systems are tabulated below:

A. Electrical and mechanical systems.

Electrical voltage system	*Electrical current system*	*Mechanical displacement system*	*Mechanical rotational system*
Voltage	Current	Force	Torque
Inductance	Capacitance	Mass	Momentum
Resistance	Conductance	Viscous friction	Viscous friction
Capacitive susceptance	Inductive susceptance	Spring stiffness	Torsional spring stiffness
Charge	Magnetic flux	Displacement	Rotation
Current	Voltage	Linear velocity	Angular velocity

B. Electrical, thermal, hydraulic and pneumatic systems

Electrical system	*Thermal system*	*Hydraulic system*	*Pneumatic system*
Charge	Heat	Liquid	Air
Current	Thermal conduction	Flow	Flow
Voltage	Temperature	Head	Pressure
Resistance	Thermal resistance	Flow friction	Flow opposition
Capacitance	Thermal capacity	Volume	Volume

4.7. ANALYSIS OF CONTROL SYSTEMS

Control system analysis work is possible after its due understanding and convenient representation. Various methods of representing a current system are :

(i) Block diagram

It is a pictorial representation of the system where the system is conveniently divided into inter-related convenient units, each unit represented by a block and connected by arrows with other blocks for its inputs and outputs. For example Fig. 4.2 is such a block diagram.

(ii) Transfer function

In reality every control system may be treated to be composed of various subsystems and every such subsystem of smaller subsystems. Thus we reach the units of the system whose understanding is easy and convenient.

Once individual units are analysed, these may be grouped together to represent the system with a few larger units and ultimately, by this process we can reach a stage where our whole system is represented by a single block with input and output.

Every unit, subsystem or system has inputs and outputs and the ratio of the two is said to be the transfer function of the unit, subsystem or system.

Writing the transf er function of smaller units is easier, hence it is done first. While grouping the units into subsystems and so on, the transfer functions may also be combined together to ultimately get a single transfer function of the whole system.

The block diagrams are of great help in this process of fragmentation and recombination. Various rules combining together the transfer functions are given in table below:

Purpose	*Unit diagram*	*Equivalent diagram*
1. Combining blocks in cascade	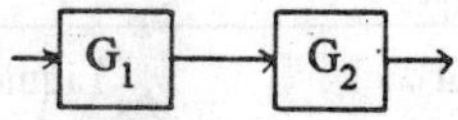	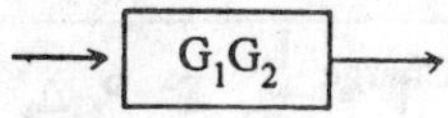
2. Moving an adder forward	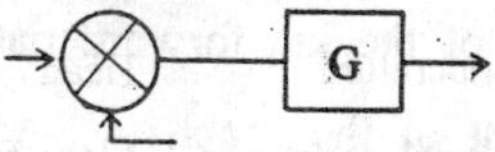	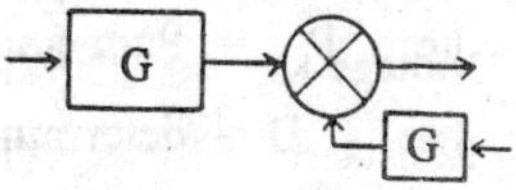
3. Moving an adder backward	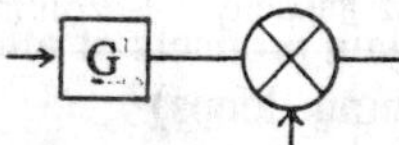	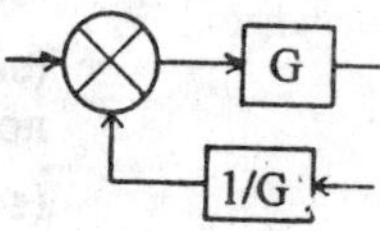
4. Moving a takeoff forward	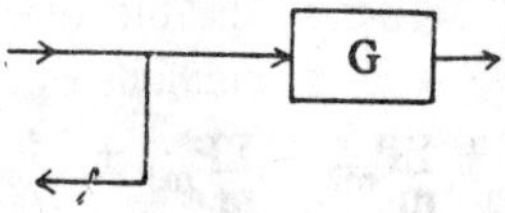	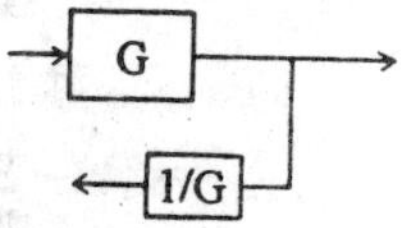
5. Moving a takeoff backward	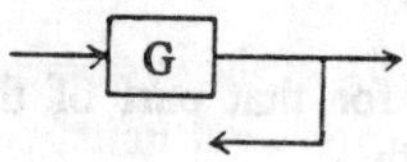	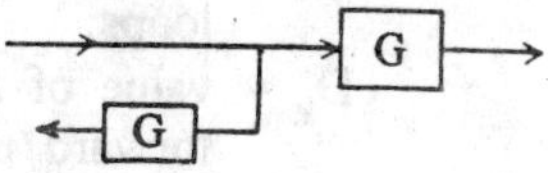
6. Eliminating a feedback	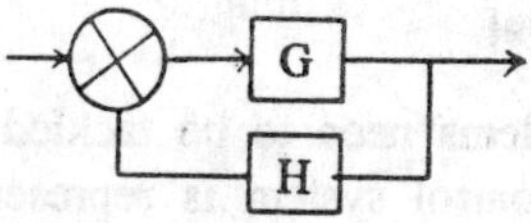	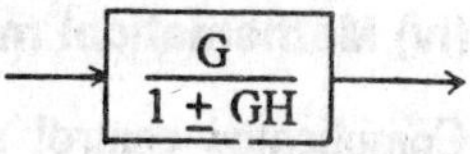
7. Combining blocks in parallel		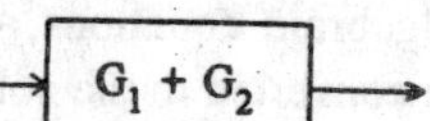

(iii) Signal flow graph

Signal flow graph is a graphical representation by the help of nodes and paths to show the relationships of variables of linear algebraic equations.

A node which has only an output path is called an input node while a node which has only an input path is known as the output node.

In the signal flow diagram nodes represent the various quantities and the paths represent the operational gain. Mason's gain formula is of great help in the analysis of a system through signal flow graph technique.

The relationship between the quantities of input and output nodes is the net gain of the system, which is given by the Mason's gain formula :

$$T = \frac{1}{\Delta} \sum_{k} P_k \Delta_k$$

where P_k = Path gain of the kth forward path.

D = determinant of the graph

= 1 – (sum of gains of all individual loops)

= + (sum of gain products of all possible combinations of two non-touching loops)

– (sum gain products of all possible combinations of three non-touching loops)

+ . . . .

$$= 1 - \sum_{m} P_{m1} + \sum_{m} P_{m2} - \sum_{m} P_{m3} + \ldots$$

where P_{mr} = gain product os mth possible combinations of r non-touching loops.

D_k = value of Δ for that part of the graph not touching the kth forward path.

(iv) Mathematical model

Complicated control systems need to be tackled mathematically. In mathematical modelling, a control system is represented by linear differential equations.

The solution of the linear differential equations is obtained by the help of Laplace transforms, which converts the differential equations into linear algebraic equations, which can be solved easily. The solution, so obtained is converted to the solution of the original equations by application of inverse Laplace transforms.

4.8. TIME RESPONSE ANALYSIS

A control system, in practice, may have various types of inputs from the simplest to the most complex. Hence defining the input and system's response to these inputs may be a tedious job, if not impossible. Hence systems are defined in terms of response to standard time-functions. This, on one hand, provides the indication for the system's response under different situations, on the other hand, it provides facility for comparative study of various systems. The standard time-functions used for control system testing are discussed below.

4.8.1. Standard test functions

(a) Unit step function

This function may be defined as :

$f(t) = 0$ for $t < 0$

$= 1$ for $t => 0$

This function due to sudden variation and steadiness later provides clear indication of system's transient response.

(b) Unit ramp function

This function is defined as :

$f(t) = k.t$ where k is a constant of value 1.

It is a constant velocity function and used to test the system's steady-state performance.

(c) Time-square or parabolic function

This is defined as:

$f(t) = kt^2$ where $k = 1$ for unit parabolic function and is a uniform acceleration function. In velocity function input, there is a steady-state error in the system's response. Acceleration function input, there is no such error.

(d) Sine function

For frequency response testing of the control systems, a sinusoidal function is the ideal input. It is represented by :

$f(t) = A \sin \omega t$

(e) Unit impulse function

A function of infinitesimally small duration Δt and of amplitude A such that :

$A.\Delta t = 1$

is called a unit impulse function.

The steady-state errors of various types of control systems for different unit test signals are given below:

Test signal	*Steady-state error*		
	Type 0 system	*Type 1 system*	*Type 2 system*
Unit step	$1/(1 + k_p)$	0	0
Unit ramp	∞	$1/k_v$	0
Unit parabolic	∞	∞	$1/k_a$

where $k_p = \lim_{s \to 0} G(s)$, $K_v = \lim_{s \to 0} SG(s)$, $K_a \lim_{s \to 0} s^2G(s)$

4.8.3. Second-order response

The systems, describable by a second-order differential equation are called second-order systems. The response of even the higher order systems is also nearly of second order. Hence study of second-order response is very important. The standard form of a second order transfer function is :

$$T(s) = \frac{w_n^2}{s^2 + 2\xi \omega_n + \omega_n^2}$$

When ξ is called the sampling ratio and ω_n is the undamped natural frequency. The roots of characteristic polynomial depend upon ξ as provided in the Table 4.2.

Table 4.2.

Range of damping ratio	*Type of response*	*Root locations*
$\xi > 1$	Overdamped	$-\xi \omega_n \pm \omega_n(\xi^2 - 1)^{½}$
$\xi = 1$	Critically damped	$-\omega_n, -\omega_n$
$0 \leq \xi < 1$	Underdamped	$-\xi \omega_n \pm j\omega_n (1 - \xi^2)^{½}$

Fig. 4.3 shows the unit-step response for a underdamped system when $0 < \xi < 1$. Under this condition, the radian frequency is given by :

$$\omega = \omega_n(1 - \xi^2)^{½}$$

called the damped natural frequency. The oscillatory response has three associated figures of merit :

(i) *Percent overshoot* : Defined by :

$$\% \text{ overshort} = 100 \times \frac{(\text{max value}) - (\text{steady-state value})}{(\text{steady-state value})}$$

(ii) *Rise time* : Defined as the time required for the unit step response to rise from 10% of the steady-state value to 90% of the steady-state value.

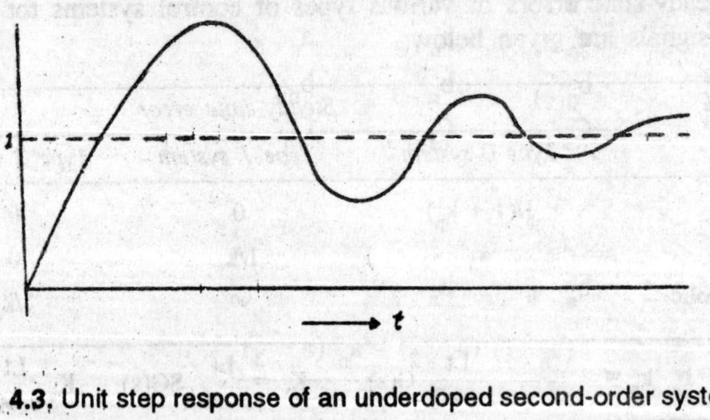

Fig. 4.3. Unit step response of an underdoped second-order system.

(iii) *Settling time* : Defined as minimum time after which the response remains within $\pm$ 5% of the steady-state value.

4.9. SYSTEM STABILITY

Consider the transfer function T(s) of a single-input, single-output closed loop system:

$$T(s) = \frac{p(s)}{q(s)}$$

where q(s) is the characteristic polynomial where roots are the poles of the system. To obtain a bounded response to a bounded input, the poles of the closed-loop system must be in the left-hand portion of the s-plane.

A necessary and sufficient condition that a feedback control system be stable is that all the poles of the system transfer function have negative real parts. The system will be unstable if all the poles are not in the left hand half plane.

If some poles lie on the imaginary axis with all other poles on the left-hand side, the system is called marginally stable.

The characteristic equation may be written as :

$$q(s) = a_n s^n{}_L + a_{n-1} s^{n-1} + \ldots + a_1 s + a_o = 0$$

The necessary but not sufficient conditions of stability are :

(i) All the coefficients of the polynomial must have the same sign.

(ii) All the coefficients be non-zero.

If any of the above condition is not satisfied, we know that the system is unstable. If these conditions are satisfied, we proceed further to test the stability.

4.9.1. Routh-Hurwitz criterion

It provides both necessary and sufficient conditions for stability. In this criterion the coefficients of the characteristic equation are ordered into an array as follows :

$$\begin{array}{lllll}
s^n & a_n & a_{n-2} & a_{n-4} & \ldots \\
s^{n-1} & a_{n-1} & a_{n-3} & a_{n-5} & \ldots \\
s^{n-2} & b_{n-1} & b_{n-3} & b_{n-5} & \ldots \\
s^{n-3} & C_{n-1} & C_{n-3} & C_{n-5} & \ldots \\
\cdot & & & & \\
\cdot & & & & \\
\cdot & & & & \\
s^o & h_{n-1} & h_{n-3} & &
\end{array}$$

where $b_{n-1} = \dfrac{(a_{n-1})(a_{n-2}) - a_n (a_{n-3})}{a_{n-1}}$

$$= -\frac{1}{a_{n-1}} \begin{bmatrix} a_n & a_{n-2} \\ a_{n-1} & a_{n-3} \end{bmatrix}$$

$$b_{n-3} = -\frac{1}{a_{n-1}} \begin{bmatrix} a_n & a_{n-4} \\ a_{n-1} & a_{n-5} \end{bmatrix}$$

and $c_{n-1} = -\frac{1}{b_{n-1}} \begin{bmatrix} a_{n-1} & a_{n-3} \\ b_{n-1} & b_{n-3} \end{bmatrix}$

and so on.

The Routh-Hurwitz criterion states that the number of roots of q(s) with positive real parts is equal to the number of changes of sign in the first column of the array. Thus for a stable system there should be no change of sign in the first column of the array. This requirement is both necessary and sufficient.

Root-Locus method

Consider a transfer function :

$$\frac{C(s)}{R(s)} = \frac{G(s)}{1 + G(s)H(s)}$$

with the characteristic equation :

$1 + G(s)\, H(s) = 0$

or $G(s)\, H(s) = -1$

Since G(s) H(s) is a complex quantity, it can be split into two equations of angles and magnitudes of both the sides.

Hence $\angle G(s)\, H(s) = \pm\, 180° \,(2k + 1)$ (where k = o, 1, 2, . . .)

and $|G(s)\, H(s)| = 1$

A plot of the points of the complex plane satisfying the angle condition alone is the root locus. The roots of the characteristic equation corresponding to a given value of the gain can be determined from the magnitude condition.

Rules for obtaining root locii

1. Rearrange the characteristic equation so that the parameter of interest appears as the multiplying factor in the form :

$$1 + \frac{K(s+z_1)\,(s+z_2)\ldots(s+z_n)}{(s+p_1)\,(s+p_2)\ldots(s+p_n)} = 0$$

2. Locate the open-loop poles and zeros in the s-plane. Open-loop zeros are zeros of G(s) H(s).
3. Find the starting points and terminating points of the root loci and find also

the number of separate root-locii. The points on the root locii corresponding to K = 0 are open loop poles.

4. Determine the root locii on the real axis by open-loop poles and zeros lying on it.
5. Determine the asymptoles of the root-locii.
6. Find the break-away and break in points.
7. Find the angles of departure or angles of arrival of root locii from the complex poles or at complex zeros.
8. Find the points where the root-locii cross the zero axis, by Routh's stability criterion or by letting s = jω in the characteristic equation, equating both the real and imiginary parts to zero separately, and solving for ω and k. ω gives the frequency at which root locii cross the imaginary axis and the value of k corresponds to a critical gain for stability.
9. From the open-loop transfer function :

$$G(s)\,H(s) = \frac{k(s^m + b\,s^{m-1} + \ldots b_m)}{s^n + 1_1, s^{n-1} + \ldots + a_n} \quad n \geq m$$

If n - m ≥ 2, a_1 is the negative sum of the roots of the equation and is independent of k. In such a case, if some of the roots move on the locus toward the left as k is increased, then other roots must move toward the right as k is increased. From this general shape of the root locii can be ascertained.

10. Determine the root locii in the broad neighbourhood of jω axis and the origin.

Frequency response methods

The steady-state response of a system to a sinusoidal input is called the frequency response. The frequency-response characteristic of a system can be obtained directly from the sinusoidal transfer function, which is obtained when s is replaced by jw in the usual transfer function.

For sinusoidal inputs :

$$[G(j\omega)] = \left(\frac{Y(j\omega)}{X(j\omega)}\right) = \text{amplitude ratio of the output and input sinusoids}$$

$$\text{and } G(j\omega) = \left(\frac{Y(j\omega)}{X(j\omega)}\right) = \text{Phase-shift of output input}$$

(i) Bode diagrams

A Bode diagram consists of two graphs, both plotted against frequency in log scale. One is a plot of log of magnitude of the sinusoidal pairs for function, and the other is a plot of the phase angle.

The standard representation of the log magnitude of G(jω) is 20 log G(jω), at base of 10 and in units of decibels. A semilog paper, with frequency on log axis, is more convenient.

Method

(i) Rewrite the sinusoidal transfer function.

(ii) Identify corner frequencies which are at $\omega = \frac{1}{T}$

(iii) Draw the asymptotic log magnitude curves.

Analysis

(a) Minimum phase systems and non-minimum phase systems - Transfer function having no poles or zeros in the right-halfs plane are minimum phase transfer functions and such systems are called minimum phase systems. If some poles and/or zeros are in right half s plane the system is called a non-minimum system.

Non-minimum phase systems are slow in response and should be avoided.

(b) System types : Control systems may be of type 0, 1 or 2.

Type 0

The first order type zero system has transfer function of the form :

$$G(jw) = \frac{K_o}{1 + j\omega T_a}$$

and will have the following characteristics:

1. The slope of frequency response at low frequencies will be zero.
2. The gain k_o is the steady-state step error coefficient.

Type 1

The second order type 1 system has the transfer function of the form :

$$G(j\omega) = \frac{k_1}{j\omega(1 + j\omega T_a)}$$

and the log frequency response will have the following characteristics :

1. The slope at low frequencies is -20dB/decade.
2. The intercept of low-frequency slope of -20dB/decade, with the 0-dB axis occurs at the frequency $\omega_n = k_1$.
3. The gain k_1 is the steady state ramp error coefficient.

Type 2

Type 2 system has a transfer function of the form :

$$G(j\omega) = \frac{k_2}{(j\omega)^2 (a + j\omega T_a)}$$

and the log frequency response will have the following characteristics :

1. The slope at low frequencies is -40 dB/decade.
2. The intercept of the low-frequency slope of -40 dB/decade with the 0-dB axis occurs at a frequency ωy where $\omega y^2 = k_2$.

(ii) The Nyquist stability test

It is a graphical method based upon the open-loop transfer function of the system. This is also called as the Nyquist theorem, which states :

The closed-loop system is stable if and only if the net number of clockwise encirclements of the points s = –1 + j0 by the Nyquist diagram of G(s) plus the number of poles of G(s) in the right half-plane is zero.

The G(s) is the loop-gain for the unity feedback system. If the Nyquist plot of the loop gain G(s) approaches the -1 point too closely, the transient response characteristics of the closed loop system are oscillatory.

5

Power Systems

Q.1 An electrical power system means (i) conversion of energy to electrical form, (ii) transportation of electrical enrgy, (iii) utilization of electrical energy, (iv) Storage of electrical energy.

(a) only (i) is true,
(b) only (i) and (ii) are true,
(c) only (i), (ii) and (iii) are true,
(d) all are true.

Q.2 Energy is available in nature in the form of (i) mechanical, (ii) electrical, (iii) chemical, (iv) thermal.

(a) all are true,
(b) only (ii) is false,
(c) only (i) and (ii) are false,
(d) only (iv) is true.

Q.3 In a magneto-hydrodynamic generator, a conductor moves in a magnetic field and the conductor is of -

(a) super-conducting material,
(b) hot ionized gas,
(c) ordinary copper conductor,
(d) none of the above three.

Q.4 The advantages of a.c. power system over d.c. power system are (i) easy transformation, (ii) simplicity of a.c. generator, (iii) lower resistance of material on a.c..

(a) only (i) is true,
(b) only first two are true,
(c) only last two are true,
(d) all are true.

Q.5 For very long distance transmission, d.c. transmission is still beneficial.

(a) true,
(b) false,
(c) nonsense,
(d) partially true

Q.6 The choice of transmission voltage ia matter of balancing between -

(a) cost of line and the cost of operation,

(b) generation and utilization voltages,
(c) generation and utilization efficiencies,
(d) power and ditribution transformations.

Q.7 At higher voltages of transmission, the energy losses increase due to -
(a) skin effect,
(b) corona effect,
(c) thermal effect,
(d) altitude effect.

Q.8 Due to power transmission system, radio interference is caused by -
(a) high frequency transmission,
(b) high voltage transmission,
(c) high current transmission,
(d) high volt drop in line.

Q.9 In power systems, the reserve capacity means -
(a) idle capacity to meet peak demands,
(b) maximum permissible capacity,
(c) capacity never utilized.
(d) capacity of generators kept floating.

Q.10 In power stations, spinning reserve means -
(a) idle capacity to meet peak demands,
(b) capacity never utilized,
(c) unused capacity of generators in operation,
(d) capacity og generators kept floating.

Q.11 Interconnection of generating stations is advantageous because it reduces the need of (i) reserve capacity, (ii) spinning reserve, (iii) limiting loads on the system, (iv) close monitoring.
(a) only (i) is true,
(b) only (i) and (ii) are true,
(c) only (iv) is false,
(d) all are true.

Q.12 Disadvantages of power system interconnection are (i) spreading of faults to the all stations, (ii) higher breaker capacities, (iii) needs of synchronization, (iv) costlier power.
(a) only (i) is true,
(b) only (i) and (ii) are true,
(c) only first three are true,
(d) all are true.

Q.13 In power system 'load study' means knowing (i) voltage, (ii) current, (iii) power, (iv) power factor, at various points of the system.
(a) these fully define the load study,
(b) some more points are missing,
(c) some points are redundant,
(d) all are wrong.

Q.14 In power system studies, a calculating board means -
(a) a scaled replica of the power system,
(b) a calculator to calculate complex vectors,
(c) a digital calculator,
(d) any thing else.

Q.15 The most economical power generation by an adjoint of alternators is obtained by -
(a) first fully loading the most economical unit,
(b) equating the cost of operations of all units,
(c) equating the increment cost of all units,
(d) any other criteria.

Q.16 Load on an alternator connected to a bus is determined by -
(a) its reactance, (b) input power,
(c) terminal voltage, (d) generated frequency.

Q.17 Most faults on transmission lines of 132 kv and above, are caused by lightenings, resulting in flashover of insulators.
(a) false, (b) partially true,
(c) true, (d) nonsense.

Q.18 Transmission line flasovers are caused by -
(a) temperature rise,
(b) ionization of surrounding air,
(c) large inductive reactance,
(d) can't say.

Q.19 Most of the transmission line flashover faults can be cleared by deenergizing the line for about -
(a) half second, (b) one hour,
(c) one day, (d) cooling time.

Q.20 Of the total transmission faults, single line-to-ground faults account for nearly -
(a) 70 - 80 percent, (b) 40 - 50 percent,
(c) 10 -12 percent, (d) zero.

Q.21 Ofthe total transmission line faults, three phase symmetrical faults account for nearly -
(a) 90 percent, (b) 60 percent,
(c) 30 percent, (d) 5 percent.

Q.22 If various alternators are connected to bus, load change on the bus is shared by various alternators depend upon (i) their input powers, (ii) impedances, (iii) their power factors (iv) terminal voltages.
(a) only first two are true, (b) only last two are true,
(c) only (i) and (iv) are true, (d) only (ii) and (iii) are true.

Q.23 For an alternator connected to a bus, stability limit means -
(a) maximum loading capacity, (b) maximum input power,
(c) minimum power factor, (d) minimum load.

Q.24 In power systems, stability study means study of conditions of -
(a) suitable loading of varios alternators,
(b) maintaining proper power factors,
(c) maintaining synchronism of various machines,
(d) anything else.

Q.25 The important parameters of a transmission line, are (i) resistance, (ii) inductance, (iii) capacitance, (iv) conductance.
(a) all are true, (b) only first three are true,
(c) only first two are true, (d) only first one is true.

Q.26 For every current-carrying conductor, fields produced by it are (i) electric field produced by the potential, (ii) magnetic field produced by the current, (iii) electric field produced by current, (iv) magnetic field produced by voltage.
(a) only first two are true, (b) only last two are true,
(c) only (ii) and (iv) are true, (d) only (i) and (iii) are true.

Q.27 In closed paths, around a current-carrying conductor, is the -
(a) magnetic field,
(b) electric field,
(c) both electric and magnetic fields,
(d) none of the two fields.

Q.28 Electric field produced by any energized conductor is -
(a) from point of higher potential to a point of lower potential,
(b) arond the conductor,
(c) fromlower potential to higher potential,
(d) nowhere.

Q.29 Inductance of a conductor is the property of the conductor relating -
(a) applied voltage and the resultant current,
(b) induced voltage and the rate of change of current,
(c) applied voltage and change of current,
(d) rate of change of flux and current.

Q.30 For a current-carrying conductor, if i is the changing current in amperes, and L is the inductance in henrys, and N the flux linkage in weber-turns and e be the induced voltage, then (i) e = dT/dt, (ii) e = Ldi/dt, (iii) L = dT/di.
(a) all are true, (b) only first two are true,
(c) only last two are true, (d) none is true.

Q.31 Transmission line conductors are made by stranding a large number of smaller diameter conductors for (i) reducing skin effect, (ii) providing flexibility to the conductor, (iii) saving on conductor material.

(a) all are true, (b) only first two are true,
(c) only first one is true, (d) none is true.

Q.32 Number of strands in a particular layer of a composite conductor of a transmission line are given by the formula $s = 3n^2 - 3n + 1$ where n is the number of layer.

(a) true, (b) false,
(c) nonsense.

Q.33 In aluminium transmission line conductors, central strand or the layer are made up of steel to provide -

(a) mechanical strength, (b) vibration resistance,
(c) higher electrical resistance, (d) economy in cost.

Q.34 Transmission line conductors made of aluminium with steel reinforcement are usually known as -

(a) ACSR, (b) ASCR,
(c) SRAC, (d) SCAR.

Q.35 A 30/19 ACSR conductor means -

(a) 30 steel and 19 aluminium strands,
(b) 30 aluminium and 19 steel strands,
(c) 30 strands in 19 layers,
(d) none of the above.

Q.36 Transmission line conductors made of copper -

(a) do not have steel reinforcement strands,
(b) have steel reinforcement strands,
(c) have reinforcement of alloy steel.

Q.37 Transmission line conductors for high voltage applications, are expanded in area by paper layers between steel and aluminium strands, to -

(a) increase dia and reduce corona,
(b) save on conductor material,
(c) make the conductor lighter,

Q.38 Geometrical mean distance (GMD) between a group of m points and another group of m points is (mn)th root of the product of (mn) distances from each point in one group to each point in the other group.

(a) true, (b) false,
(c) partially true, (d) nonsense.

Q.39 Self-GMD of a composite conductor having n strands, also known as GMR (geometric mean radius), is the n2 root of product of distances from each of the strands to itself and to all other strands.

(a) true, (b) false,
(c) partially true, (d) nonsense.

Q.40 For a two-conductor line having composite conductors, if D_m and D_s be the GMD and GMR for a conductor, thr inductance of the conductor is given by $L = 2 \times 10^{-7} \ln D_m/D_s$ henrys/meter

(a) false, (b) true,
(c) partially true, (d) nonsense.

Q.41 Self-GMD of a circular conductor, having a radius R, is given by -

(a) 0.7788xR, (b) $0.7xR^2$,
(c) R, (d) something else.

Q.42 GMD between two circular conductors, with distance D between the two centers, is given by -

(a) D^2, (b) D,
(c) D/2, (d) something else.

Q.43 GMD from an external point to a circular conductor, withdistance D between the point andthe center of the conductor, is given by -

(a) D, (b) D/2,
(c) D^2, (d) something else.

Q.44 GMD between two annular rings, with centers D distance apart, is given by -

(a) D, (b) D^2,
(c) D/2, (d) something else.

Q.45 Self-GMD (GMR) of a rectangular area of AxB dimensions, is given by -

(a) 0.2235(A + B), (b) 0.2235 AB,
(c) $0.2235\ (AB)^2$, (d) something else.

Q.46 Inductance of a composite conductor of a three-phase equilaterally-spaced transmission line, with distance D between the centers of the conductors and with self-GMD of each conductor D_s, is given by $L = 2 \times 10^{-7} \ln (D/D_s)$ henrys per meter.

(a) true, (b) false,
(c) partially true, (d) nonsense.

Q.47 If the three identical conductors of a transmission line are not symmetrically spaced, their -

(a) inductances will be unequal,
(b) resistances will be unequal
(c) their temperature rise will beunequal,
(d) inductances willbe equal.

Q.48 Conductors of a long three-phase transmission line are transposed, to equate their -

(a) inductances, (b) resistances,
(c) currents, (d) voltages.

Q.49 A transmission line is said to have bundled-conductors, if its each phase has (i) more than one conductors, (ii) conductors of a phase run parallel, (iii) conductors of a phase are closely spaced.

(a) all are true, (b) only first two are true,
(c) only first one is true, (d) none is true.

Q.50 As compared to a composite conductor, a bundled conductor has (i) lower reactance, (ii) lower voltage gadient, (iii) reduced radio interference.

(a) all are true, (b) only first two are true,
(c) only first one is true, (d) all are false.

Q.51 Optimum spacing between the conductors of a bundled-conductor is -

(a) 8-10 times the conductor diameter,
(b) independent of conductor diameter,
(c) inverse of conductor diameter,
(d) none of the above.

Q.52 Capacitance between parallel conductors of a transmission line is (i) charge per unit of potential difference between the conductors, (ii) dependent upon size of the conductors, (iii) dependent upon spacing between the conductors.

(a) all are true,
(b) only first two are true,
(c) only last two are true,
(d) all are false.

Q.53 In an open-circuited A.C. transmission line, charging current between the conductors, is produced because of -

(a) the alternating potential difference,
(b) the inductances of the conductors,
(c) the resistances of the conductors,
(d) some other reason.

Q.54 Charging current of a transmission line (i) increases the volt drop in the line, (ii) decreases the efficiency, (iii) improves the pwer factor, (iv) affects the stability of the power system.

(a) all are true,.
(b) all except (ii) are true,
(c) all except (iii) are true,
(d) all except (iv) are true.

Q.55 Electric flux density is the electric flux per square meter and is measured in -

(a) webers, (b) coulombs,
(c) farads, (d) ohms

Q.56 Electric flux is caused by potential difference and flows -

(a) around the conductor,
(b) from point of higher potential to that of lower potential,
(c) from conductor to ground,
(d) along the conductor.

Q.57 Charge on a conductor is -

(a) distributed on its surface,
(b) concentrated at its center,
(c) uniformly distributed in the space around,
(d) non-existent.

Q.58 All equidistant points from a charged conductor have (i) equal potential, (ii) equal charge, (iii) equal electric flux density.

(a) all are true, (b) all except (i) are true,
(c) all except (ii) are true, (d) all except (iii) are true.

Q.59 Electric field intensity at a point away from a charged point is -

(a) negative of the potential gradient,
(b) the potential gradient,
(c) charge density,
(d) none of the above.

Q.60 Electric field intensity at a point away from a charged conductor, is equal to -

(a) electric field density/permitivity of the medium,
(b) potential at that point,
(c) magnetic field intensity,
(d) none of theabove.

Q.61 The effect of earth on the capacitance of a transmission line can be studied by an image of the line -

(a) below the earth surface,

(b) on the earth surface,
(c) parallel to the line and above the earth surface.
(d) none of the above.

Q.62 The image of a transmission line conductor, is assumed to have -
(a) charge equal to that of the conductor,
(b) charge equal and opposite to that of the conductor,
(c) potential equal to that of the conductor,
(d) none of the above.

Q.63 The concept of GMD (geometric mean diameter), as used for transmission line conductors inductance calculations, can be applied to capacitance calculations also -
(a) in original form, (b) in modified form,
(c) not applicable, (d) nonsense.

Q.64 The GMR (geometric mean radius) of a conductor, as applied in inductance calculations for a transmission line, has to be replaced by actual radius of the conductor, for capacitance calculations because -
(a) the charge resides on the surface of the conductor,
(b) the charge is centered,
(c) electric flux lines are concentric,
(d) none of the above.

Q.65 Modified-GMD of a transmission line, means -
(a) GMR replaced by radius of the conductor,
(b) GMR replaced by self-GMD of the conductor,
(c) transposed GMD,
(d) none of the above.

Q.66 Capacitive susceptance means -
(a) reciprocal of capacitive reactance,
(b) reciprocal of capacitance,
(c) negative of capacitive reactance,
(d) none of the above.

Q.67 Effective resistance of a conductor means -
(a) volt drop / current,
(b) power loss / square of current,
(c) 90 percent of resistance,
(d) d.c. resistance.

Q.68 As compared to the d.c. resistance, the a.c. resistance of a solid conductor is mostly -

(a) more, (b) less,
(c) equal,

Q.69 The stranding of a conductor for transmission line conductor (except for the central conductor) increases the length of the conductor by nearly -
(a) 0.5 percent, (b) 2 percent,
(c) 5 percent, (d) 10 percent.

Q.70 Permeability of space is -
(a) $4\pi \times 10^{-7}$ henry/meter, (b) $4\pi \times 10^{7}$,
(c) 4.3, (d) none of above.

Q.71 Corona losses on a transmission line increase as the (i) voltage gradient, (ii) diameter of the conductor, (iii) spacing between conductors, (iv) ambience dampness, increase.
(a) all are true, (b) all except (i) are true,
(c) all except (ii) are true, (d) only (ii) and (iii) are false.

Q.72 Voltage regulation of a transmission line is rise in voltage at the receiving-end, expressed in percentage of full-load voltage, when full load, at the specified power factor, is removed while the sending-end voltage is held constant.
(a) correct, (b) mistaken,
(c) partially true, (d) nonsense.

Q.73 In the definition of voltage regulation of a transmission line, scalar values of the voltages are considered.
(a) true, (b) false,
(c) partially true, (d) nonsense.

Q.74 The voltage regulation of a transmission line is measured at the -
(a) receiving end, (b) sending end,
(c) both ends, (d) at the centre.

Q.75 The voltage regulation of a transmission line is referred -
(a) to a specific load power factor,
(b) irrespective of the power factor,
(c) as same for all power factors,

Q.76 For a given receiving-end voltage, in a transmission line, sending-end voltage is highest for -
(a) inductive load, (b) capacitive load,
(c) resistive load, (d) no load.

Q.77 In a long transmission line, for leading loads,
(a) sending-end voltage > receiving-end voltage,

(b) reverse of (a),
(c) sending-end voltage = receiving-end voltage,

Q.78 In short transmission lines calculations, usually -
(a) capacitance is ignored, (b) inductance is ignored,
(c) resistance is ignored. (d) voltage drop is ignored.

Q.79 In medium-length transmission lines calculations, usually (i) nominal-T, (ii) nominal-¶ circuit, is assumed.
(a) both are true, (b) only (i) is true,
(c) only (ii) is true, (d) noneis true.

Q.80 For a transmission line, if z is the series-impedance per unit length and y the shunt-admittance per unit length, its characteristic impedance is -
(a) √(z/y), (b) z/y,
(c) yz, (d) √(yz).

Q.81 For a transmission line, if z and y are respectively the series-impedance and the shunt-admittance per unit length of the line, its propagation constant is given by -
(a) √(z/y), (b) z/y,
(c) yz, (d) √(yz).

Q.82 Attenuation constant in nepers per unit length of a transmission line is -
(a) real part of propagation constant,
(b) imaginary part of propagation constant,
(c) real part of characteristic impedance,
(d) imaginary part of characteristic impedance.

Q.83 Phase constant in radians per unit length of a transmission line is -
(a) real part of propagation constant,
(b) imaginary part of propagation constant,
(c) real part of characteristic impedance,
(d) imaginary part of characteristic impedance.

Q.84 In a long transmission line, voltages and currents can be regarded as waves and at any point, the net voltage or current is the -
(a) sum of incident and reflected waves,
(b) difference of incident and reflected waves,
(c) either incident or reflected wave,
(d) none of the above.

Q.85 If a transmission line is terminated at the receiving-end by its characteristic impedance, the line is called a flat or infinite line because -

(a) reflected wave will be equal to incident wave,
(b) reflected wave will be zero,
(c) incident wave will be zero,
(d) none of the above.

Q.86 The communication lines are usually made flat or infinite by terminating by -

(a) characteristic impedance, (b) propagation impedance,
(c) fixed resistance, (d) none ofthe above.

Q.87 For a long transmission line, of single circuit the typical value of its characteristic impedance is -

(a) 400 ohms, (b) 200 ohms,
(c) 100 ohms, (d) 800 ohms.

Q.88 A transmission line is called a lossless line if its characteristic impedance is equal to surge impedance ie -

(a) $\sqrt{(L/C)}$, (b) L/C
(c) $\sqrt{(LC)}$, (d) none of above.

Q.89 The term 'Surge Impedance Loading (SIL)' of a transmission line is used to deal with lightening surges or high frequencies, and under such conditions -

(a) line losses are neglected,
(b) line losses are very important,
(c) line resistance become dominant,
(d) none of the above.

Q.90 Surge Impedance Loading (SIL) of a transmission line is the power delivered by it to a purely resistive load equal to its surge impedance.

(a) true, (b) false,
(c) partially true, (d) nonsense.

Q.91 A transmission line is (i) passive, (ii) bilateral, (iii) linear, (iv) two-terminal pair, network.

(a) all are true, (b) only first two are true,
(c) only last two are true, (d) only (ii) and (iii) are true.

Q.92 For a transmission line if V_s an I_s are sending-end voltage and currents and V_r and I_r receiving-end voltage and current, and A, B, C, D are generalized circuit constants -

(a) $V_r = D.V_s - B.I_s$, (b) $V_s = A.V_r - B.I_r$,
(c) $I_s = C.V_r - D.I_r$, (d) $I_r = C.V_s - A.I_s$.

Q.93 For a transmission line if V_s an I_s are sending-end voltage and currents and V_r and I_r receiving-end voltage and current, and A, B, C, D are generalized circuit constants -

(a) AB - CD = 0, (b) AD - BC = 1,
(c) AD - BC = 0, (d) AB - CD = 1.

Q.94 For a short transmission line when the shunt admittance is neglected A = 1, B = Z, C = 0, D = 1, where Z is series impedance.

(a) true, (b) false,
(c) nonsense,

Q.95 For all overhead transmission lines, usually -

(a) L x C is constant, (b) L / C is constant,
(c) L + C is constant, (d) L - C is constant.

Q.96 A resistance or a reactor in the neutral grounding, limits the current during -

(a) normal operation, (b) ground fault,
(c) phase short-circuit fault, (d) resonance.

Q.97 A Petersen coil is connected -

(a) in alternator neutral grounding,
(b) in all transformer neutral grounding,
(c) in transformer neutral grounding below 70 kv,
(d) in lightening arresters.

Q.98 In power system fault calculations, synchronous motor loads connected to the system are always taken into consideration because -

(a) their torques continue to draw current,
(b) their EMFs contribute to fault current,
(c) they are major loads on the system,
(d) their fields may get over-excited.

Q.99 In power system fault calculations, induction motor loads affect the fault current -

(a) for a few minutes, (b) for first few cycles only,
(c) never, (d) always.

Q.100 In power system calculations, transformer windings are referred to by their -

(a) actual impedances, (b) equivalent values,
(c) subtransient values, (d) steady state values,

Q.101 For power system calculations, if the diagram refers to the voltage different from that generated, generator reactances are referred by -

(a) actual reactances, (b) equivalent reactances,
(c) transient reactances, (d) steady-state values.

Q.102 The equivalent reactances of the different voltage sides are obtained by multiplying the actual reactances by -
(a) transformation ratio,
(b) square of the transformation ratio,
(c) generated voltage,
(d) 2.

Q.103 In power system calculations, generator and transformer voltages are referred by -
(a) actual voltages, (b) equivalent voltages,
(c) transient values, (d) lowest values.

Q.104 In power system computations, the equivalent voltages of the different voltage side, are obtained by multiplying the actual voltages by -
(a) transformation ratio,
(b) square of the transformation ratio,
(c) generated voltage,
(d) 2.

Q.105 For power system computations, to refer voltages, currents and impedances by per unit or percent of a selected base, the base values selected are -
(a) any values,
(b) lowest values of the system,
(c) highest values of the system,
(d) any steady-state value.

Q.106 Per unit value of any quantity in a power system is defined as the ratio of the quantity to -
(a) base value in decimal notation,
(b) its inverse,
(c) base value in fractions,
(d) 100.

Q.107 The four base values for voltage, current, power and impedance, for a particular power system, are selected -
(a) arbitrarily, (b) inter-relatively,
(c) independently, (d) to be the same.

Q.108 Out of the four base values for voltage, current, impedance and power,
(a) any two are selected and others are computed,
(b) all four are selected,

(c) all four are computed,
(d) any of the above method can be adopted.

Q.109 If F is input cost for a given load P, the increment fuel cost for the power generating unit for that power output, is -
(a) F/P, (b) dF/dP,
(c) F/dP, (d) dF/P.

Q.110 For generation of power from a multi-unit power station, with minimum cost of operation, units must be loaded -
(a) in ratio of their capacities, (b) equally,
(c) with equal increment cost, (d) none of the above.

Q.111 In an alternator, transient fault currents are different from that at steady state because of -
(a) the delay of tripping devices,
(b) the armature reaction,
(c) the delay in field suppression,
(d) the switching surges.

Q.112 The selection of a circuit breaker in a power system depends upon (i) load on the breaker, (ii) fault current, (iii) switching surges.
(a) only (i) is true, (b) only (i) and (ii) are true,
(c) only (i) and (iii) are true, (d) all are true.

Q.113 Momentary rating of a circuit breaker should be around -
(a) the switching surge current,
(b) 1.5 times the subtransient maximum fault current,
(c) twice the load current,
(d) all theabove.

Q.114 In an R-L circuit, if an a.c. voltage is applied, usually, the ensuing current has the following two components -
(a) one a.c. and the other d.c., (b) both a.c.,
(c) both d.c., (d) none of the above.

Q.115 The transient component of current in an R-L circuit, due to switching-on an a.c. voltage in it, can be eliminated by -
(a) resonance in the load,
(b) switching at a particular moment of the voltage wave.
(c) a certain R/L ratio.
(d) all of the above.

Q.116 The initial current in an R-L circuit due to an a.c. voltage depends upon (i) value of the voltage at the instant of switching, (ii) power factor of the load, (iii) r.m.s. value of voltage, (iv) load imedance.

(a) only first one is true, (b) only first two are true,
(c) only last two are true, (d) all are true.

Q.117 Synchronous reactance of an alternator is -

(a) open-circuit voltage/steady-state current at short-circuit,
(b) armature reactance at synchronous speed,
(c) both of above,
(d) none of above.

Q.118 Subtransient current due to a fault in an alternator is -

(a) initial current,
(b) initial symmetrical r.m.s. current.
(c) initial peak current,
(d) none of the above.

Q.119 For an alternator, reactances -

(a) have fixed values,
(b) are dependent upon saturation,
(c) are dependent upon excitation,
(d) totally fictitious.

Q.120 In power system studies, stability of a machine, means -

(a) its capability to maintain synchronism with the system,
(b) its steadiness of speed,
(c) its steadiness of load,
(d) all the above.

Q.121 In power system ircuit breaker capacity computations, (i) subtransient reactance is used for alternators, (ii) transient reactance is used for synchronous motors.

(a) both the above are true,
(b) both the above have to be reversed,
(c) only transient reactances are consudered,
(d) only subtransient reactances are considered.

Q.122 Stability limit of a power system, means -

(a) maximum disturbance it can tolerate,
(b) maximum power flow with stability,
(c) both of above,
(d) none of above.

Q.123 Steady-state stability limit of a power system is maximum flow of power possible without losing stability, when -

(a) power is maintained constant,
(b) power is increased very gradually,
(c) excitation is suddenly altered,
(d) excitation is maintained constant.

Q.124 Transient stability limit of a power system is maximum flow of power possible without losing stability, when the power is changed suddenly.

(a) true, (b) false,
(c) partially true, (d) nonsense.

Q.125 Subtransient stability limit of a power system is maximum flow of power possible without losing stability, when the power is changed suddenly.

(a) true, (b) false,
(c) partially true, (d) nonsense.

Q.126 For transmitting a given amount of power with fixed transmission efficiency, the volume of conductor material used will be proportional to -

(a) volt-ampere, (b) voltage,
(c) current, (d) wattage,

Q.127 The skin effect is directly proportional to -

(a) voltage, (b) current,
(c) frequency, (d) power factor,

Q.128 Assuming the efficiency of transmission to be fixed, the conductor cross-sectional area will be -

(a) inversely proportional to voltage,
(b) directly proportional to voltage,
(c) inversely proportional to the square of voltage,
(d) directly proportional to the square of the voltage.

Q.129 Other things except the voltage, remaining fixed, the voltage regulation of a transmission line is -

(a) directly proportional to voltage,
(b) constant,
(c) inversely proportional to voltage,
(d) none of the above.

Q.130 With a long transmission line, the best location of a synchronous condenser for power factor improvement, is -

(a) receiving-end, (b) sending-end,
(c) middle of the line, (d) anywhere.

Q.131 The synchronizing power of an alternator is enhanced by the presence of -

(a) reactance, (b) resistance,
(c) reluctance, (d) impedance.

Q.132 The synchronizing power of an alternator will be maximum when its resistance is -

(a) zero, (b) equal to leakage reactance,
(c) less than leakage reactance, (d) more than leakage reactance.

Q.133 Voltage responses of an alternator to a change in excitation, when it feeds a bus and when it feeds an isolated load, are -
(a) same, (b) different,

Q.134 For many alternators in parallel, the load sharing will be dependent upon -
(a) reactances, (b) rotor angular positions,
(c) voltages, (d) capacities,

Q.135 In a transmission line, shunt conductance is caused by leakage through -
(a) insulators, (b) air,
(c) ground, (d) all the above.

Q.136 The dielectric strength of air at normal pressure and 25°C is -
(a) 50 kv per meter, (b) 21.1 kv r.m.s. per cm,
(c) 21.1 kv r.m.s. per mm, (d) 21.1 kv peak per cm.

Q.137 High tension cables can be used upto -
(a) 110 kv, (b) 220 kv,
(c) 330 kv (d) 400 kv.

Q.138 Charging current in the cables -
(a) lags the voltage by 90°, (b) lags the voltage by -90°,
(c) lead the voltage by 180°, (d) none of the above.

Q.139 Dielectric losses in a cable increase with the increase of -
(a) load, (b) temperature,
(c) power factor, (d) all the above.

Q.140 The generalized constant A of a transmission line has the dimension of -
(a) null, (b) mho,
(c) ohm, (d) not fixed.

Q.141 The generalized constant D of a transmission line has the dimension -
(a) ohm, (b) mho,
(c) henry, (d) none of the above,

Q.142 The shunt capacitance is neglected in case of a -
(a) short transmission line, (b) long transmission line,
(c) medium short line, (d) all the above.

Q.143 For a medium transmission line, the shunt capacitance is assumed to be -
(a) lumped at the center, (b) lumped at the two ends,

(c) as (a) and (b), (d) as (a) or (b).

Q.144 By the use of transformers, the cost of transmission is usually -
(a) reduced, (b) increased,
(c) kept same, (d) nonsense.

Q.145 The whole line performance can be determined by -
(a) sending-end circle diagram,
(b) receiving-end circle diagram,
(c) universal circle diagram,
(d) all the above.

Q.146 Equal area criterion of stability is not applicable to -
(a) single machine system, (b) two machine system,
(c) multi-machine system, (d) all the above.

Q.147 Size of a feeder is determined mainly by -
(a) load current, (b) length of transmission,
(c) voltage, (d) none of the above.

Q.148 For a fixed amount of power transmission, if voltage is increased by n times, the current will -
(a) increse by n times, (b) decrease by n-square times,
(c) decrease by n times, (d) none of theabove.

Q.149 In a 3-phase four-wire system, the neutral conductor size is -
(a) equal to the phase conductors,
(b) half the other conductors,
(c) quarter of the phase conductors,
(d) none of the above.

Q.150 As for as voltage regulation is concerned, ring main system -
(a) is better than the radial system,
(b) is worse than the radial system,
(c) is same as radial system,

Q.151 Arc voltage produced in a circuit breaker is always -
(a) in phase with arc current, (b) lagging the arc current,
(c) leading the arc current,

Q.152 Discrimination of a protective device is its property to -
(a) detect a faulty condition,
(b) respond to a faulty condition,
(c) detect and respond to a faulty condition,
(d) none of the above.

Q.153 Stability of a protective device is its property -
(a) to remain stable under fault conditions,
(b) to remain stable except under the specified fault condition,

(c) to bring the system back to stability after fault clearance,
(d) none of the above.

Q.154 Sensitivity of a protective device is defined as -
(a) least fault current necessary to operate it,
(b) ratio between its operating current and normal current,
(c) function of volt-ampere input necessary to operate it,
(d) none of the above.

Q.155 Bias in a protective relay means a counter-torque in the relay -
(a) for stopping its operation under the conditions it is not supposed to operate,
(b) for operating it under the conditions of fault,
(c) a voltage across its operating coil,
(d) none of the above.

Q.156 Solid earthing of neutral in the three-phase power systems has -
(a) simplified the protective system,
(b) complicated the protective systems,
(c) not affected the protective systems,
(d) none of the above.

Q.157 As compared to impedance earthing, solid earthing of the neutral of a power system -
(a) reduces the shocks on the system,
(b) increases the shocks on the system,
(c) the shock on the system are not affected,
(d) none of the above.

Q.158 The main advantage of resistance earthing of a power system neutral is -
(a) operation of protective system,
(b) enhancement of protective system sensitivity,
(c) limitation of fault current,
(d) none of the above.

Q.159 In peterson coil earthing of power system neutral, the inductance of the coil is chosen so as to -
(a) limit the fault current to twice the full load current,
(b) getting a voltage drop sufficient to operate the relay system,
(c) balance the total capacitive current of the power system,
(d) none of the above.

Q.160 A negative-sequence relay of the protective system is that which responds to -
(a) only to balanced phase currents,

(b) only to fault currents,
(c) only to unbalanced phase currents,
(d) none of the above.

Q.161 The most important objective of a protective system is to -
(a) safeguard against the damage to the system,
(b) safety of the protective system,
(c) maintain continuity of supply,
(d) none of the above.

Q.162 The most important external factor on which the efficacy of the protective system depends upon is -
(a) soundness of the power system,
(b) interconnections of the power system,
(c) the speed of operation of thecircuit breakers,
(d) none of the above.

Q.163 To maintain continuity of supply, it is essential that -
(a) generating capacity is large enough,
(b) the power system network is well protected,
(c) each point be fed by two feeders,
(d) none of the above.

Q.164 The main advantage of radial feeders in a power system is that -
(a) supply can be maintained without interruption,
(b) the maximum fault current power is less,
(c) can be fast restored to normalacy after the fault,
(d) none of the above.

Q.165 In power systems parallel feeders are used -
(a) radial distribution system, (b) ring distribution system,
(c) in both the above, (d) none of the above.

Q.166 The more economical way of maintaing 100 percent availablity of power, through alternative feeds is -
(a) ring mains distribution system,
(b) radial distribution system,
(c) parallel feeder system,
(d) none of the above.

Q.167 In contrast to parallel feeders, split-conductor feeders are -
(a) rated half the current capacity,
(b) fed from the same circuit breaker,
(c) have no technical difference,
(d) none of the above.

Q.168 A protective system is called a qualitative type if it responds to -
(a) the measurements of the quantities within the system,
(b) the comparison of input and output of a system,
(c) comparison of the parameters of the system with standards,
(d) none of the above.

Q.169 A protective system is called a quantitative type if it responds to -
(a) the measurements of the quantities within the system,
(b) the comparison of input and output of a system,
(c) comparison of the parameters of the system with standards,
(d) none of the above.

Q.170 In qualitative protective system, running of a signal pilot wire for the whole system is -
(a) essential, (b) desirable,
(c) not necessary, (d) none of the above.

Q.171 A circuit breaker is called a high-speed type, if its total break time is -
(a) equal or less than 0.1 second,
(b) equal or less than 0.04 seconds,
(c) equal or less than 0.2 seconds,
(d) none of the above.

Q.172 The definite-time relays used in protective systems are those in which the operating time -
(a) has a definite ratio with operating current,
(b) is independent of the operating current,
(c) is independent of the high values of operating current,
(d) none of the above.

Q.173 The definite-time relays used in protective systems are of -
(a) induction type with single operating coil,
(b) induction type with three operating coils,
(c) electromagnetic type with one operating coil,
(d) none of the above.

Q.174 The definite-minimum inverse-time relays used in power system protection schemes have time delay of operation -
(a) proportional to operating current,
(b) inversely proportional to the operating current,
(c) inversely proportional to the applied voltage,
(d) none of the above.

Q.175 In nondirectional type of definite-minimum inverse-time relays used in power system protection, there is -
(a) minimum operating current and minimum operating time,
(b) maximum operating current and minimum operating time,
(c) maximum operating current and maximum operating time,
(d) none of the above.

Q.176 The directional type of definite-minimum inverse-time relays used in power system protection are of -
(a) induction type with single operating coil,
(b) induction type with three operating coils,
(c) electromagnetic type with one operating coil,
(d) none of the above.

Q.177 The directional type relays used in power system protection, the special feature is its operation with -
(a) voltage in one direction only,
(b) machine rotation in one direction only,
(c) current in one direction only,
(d) none of the above.

Q.178 The nondirectional relays used for power system protection can be used with -
(a) radial feeders only, (b) ring feeders only,
(c) parallel feeders only, (d) none of the above.

Q.179 Directional relays used for power system protection clear -
(a) the farthest fault faster,
(b) the nearest fault faster,
(c) the faults in time independent of its distance,
(d) none of the above.

Q.180 For effective protection of ring main and parallel feeders, the following type of relay is appropriate -
(a) definite-time relays,
(b) nondirectional definite-minimum inverse-time relays,
(c) directional definite-minimum inverse-time relays,
(d) none of the above.

Q.181 Directional relays used for power system protection have -
(a) only one operating coil,
(b) one current and one voltage coil for operation,
(c) one current coil and one wattage-operated element,
(d) none of the above.

Q.182 Directional overcurrent relays have one current and one watt-

operated element, but operate almost independent of voltage and power factor, because -

(a) the wattmeter element works only as an interlock,
(b) the voltage coil is made very powerful,
(c) the wattmeter element is usually bypassed,
(d) none of the above.

Q.183 In a directional overcurrent relay, the response to power flow in one direction only is achieved because the wattmeter element -

(a) responds to one direction of power,
(b) provides the output signal when power flow is in one direction,
(c) allows the operation of current element when the power flow is in one direction,
(d) none of the above.

Q.184 In a directional overcurrent relay used for power system protection, thecircuit breaker is tripped by the -

(a) current element contact only,
(b) current and wattmeter contact in series,
(c) current and wattmeter contacts in parallel,
(d) wattmeter contact only.

Q.185 As compared to the radial power distribution network, the number of circuit breakers in a ring type of network is -

(a) half, (b) same,
(c) double, (d) none of the above.

Q.186 For the protection of a ring mains power system, thc direction relay is set for power flow -

(a) from substation bus into the protected feeedr,
(b) to the substation bus from the protected feeder,
(c) on one side as in (a) and on the other side as in (b) above,
(d) none of the above.

Q.187 For a particular earth fault in a ring-main power system protected by two directio relays at each substation -

(a) all protective relays are operated,
(b) only half of the relays are operated,
(c) relays on the shorter side of the fault are operated,
(d) none of the above.

Q.188 For the protection of ring and parallel type of power systems, the connections are made such that for the unity power factor at the protected feeder,-

(a) the current leads the voltage by 30 degrees,
(b) the current and voltage are in same phase,

(c) the current leads the voltage by 45 degrees,
(d) none of the above.

Q.189 The thirty degree connection of the direction relays in a ring feeder protection, is done to keep the relay -
(a) slow-operating for low faults,
(b) torque maximum for fault conditions,
(c) torque sufficient for the fault conditions,
(d) none of the above.

Q.190 In three-phase systems, the thirty-degree connection is achieved by connecting -
(a) current coils in star and voltage coils in delta,
(b) current coils in delta and voltage coils in star,
(c) capacitors in the current coil circuit,
(d) none of the above.

Q.191 The earth-fault relay in the protection system, is operated by -
(a) positive-sequence currents, (b) negative-sequence currents,
(c) zero-sequence currents, (d) none of the above.

Q.192 A three-phase balanced earth fault does not operate -
(a) an overcurrent relay, (b) a nondirectional relay,
(c) an earth-fault relay, (d) none of the above.

Q.193 In a power system protection scheme, earth-fault relays are provided -
(a) one for each phase, (b) one for each pair of phases,
(c) one for all three phases, (d) none of the above.

Q.194 The earth-fault relays in power system protection scheme for ring and parallel feeders, are provided with -
(a) directional properties, (b) no directional properties,
(c) bidirectional properties, (d) none of the above.

Q.195 The voltage coil of an earth-fault relay in the protection of a ring feeder is fed from -
(a) a phase voltage leading the current coil by 30 degrees,
(b) a line voltage leading the current by 45 degrees,
(c) residual voltage,
(d) none of the above.

Q.196 The neutral current transformer ratio used to feed an earth-fault relay is usually -
(a) 100/5, (b) 50/5,
(c) 10/5 (d) 1/1..

Q.197 The residual voltage of a 3-phase power system, used for earth-fault relays, means -

(a) algebraic sum of the three phase voltages,
(b) zero-sequence voltage,
(c) shift of neutral point,
(d) none of the above.

Q.198 The residual voltage of a 3-phase system is dependent upon -

(a) neutral to grond impedance, (b) phase fault impedance,
(c) both of above (a) and (b), (d) none of the above.

Q.199 The residual voltage of a 3-phase system is (i) reduced by lower neutral-to-ground impedance, (ii) incresed by lower phase-to-ground fault.

(a) all the above are true, (b) only (i) is true,
(c) only (ii) is true, (d) none of the above is true.

Q.200 In time-graded protective systems, operating times of the various relays in the system, are -

(a) kept same,
(b) graded in accordance with their distance from the source,
(c) graded in accordance with their fault possibilities,
(d) none of the above.

Q.201 In distance-type protection systems, the operating times of the various relays in the system, are -

(a) kept same,
(b) graded in accordance with their distance from the source,
(c) graded in accordance with their fault possibilities,
(d) none of the above.

Q.202 The distance-type protection systems are applicable to -

(a) short feeders, (b) medium-length feeders,
(c) long-distance feeders, (d) none of the above.

Q.203 The operation of relays in a distance-relaying protection system, is based upon -

(a) voltage-drop between the relay and the fault,
(b) time taken by the fault current reaching the relay,
(c) imedance or reactance between the fault and relay points,
(d) none of the above.

Q.204 For the operation of relays in a distance-relaying protection system, dependance upon reactance is preferred to -

(a) circumvent the resistance of the line,
(b) make the relaying frequency dependent,

(c) circumvent the arc resistance of the fault,
(d) none of the above.

Q.205 The operating time of distance-time relays used for the protection of the power systems, is -
(a) fixed,
(b) proportional to the distance of the fault,
(c) inversely proportional to the fault distance,
(d) none of the above.

Q.206 The definite-distance relays used for the protection of power systems, operate -
(a) for faults within the specified distance only,
(b) within a fixed time for all faults,
(c) with delay proportional to the distance of the fault,
(d) none of the above.

Q.207 The distance-type protection relays operate from -
(a) current proportional to the fault current,
(b) voltage proportional to the voltage drop in the line upto the fault,
(c) both (a) and (b) above,
(d) none of the above.

Q.208 Distance-relaying protection system relays, have -
(a) directional element,
(b) no directional element,
(c) suppressed directional element,
(d) none of the above.

Q.209 Pilotless differential type of power system protection scheme is applicable to -
(a) ring distribution, (b) parallel feeders,
(c) radial feeders, (d) none of the above.

Q.210 Pilotless differential type of power system protection scheme operate by the difference between -
(a) the currents at the sending and receiving ends,
(b) the voltage at the sending and receiving ends,
(c) the currents in the two feeders,
(d) none of the above.

Q.211 Pilotless differential type of relays used in power system protection have two coils, one for operation and the other for biasing. The coils two are fed from -
(a) voltages of the two feeders,
(b) currents of the two feeders,

(c) the currents at the receiving and sending ends,
(d) none of the above.

Q.212 Pilotlss-differential type of power system protection scheme operate -
(a) when the two feeders carry unequal currents,
(b) when the two feeders have unequal voltages,
(c) when a feeder has inequality at the two ends,
(d) none of the above.

Q.213 Translay system of pilotless differential protection, which operates on opposed voltage basis, is applicable to -
(a) two feeders only,
(b) upto three feeders only,
(c) any number of parallel feeders,
(d) none of the above.

Q.214 Pilot-wire Lock-in protection system is used for -
(a) short feeders, (b) medium-length feeders,
(c) long-distance feeders, (d) none of the above.

Q.215 Pilot-wire Lock-in protection system operates on the basis of -
(a) amounts of power at the two ends of a feeder,
(b) direction of power at the two ends of a feeder,
(c) voltages at the two ends of a feeder,
(d) none of the above.

Q.216 Pilot-wire Lock-in protection system remain active -
(a) all the time, (b) only at the fault condition,
(c) only at the normal condition, (d) none of the above.

Q.217 In pilot-wire Lock-in protection system, under normal conditions, the receiving-end condition locks in the -
(a) sending end breaker, (b) receiving end breaker,
(c) both the breakers, (d) none of the above.

Q.218 Pilot-wire Lock-in protection system, under fault conditions, the first to operate is -
(a) sending-end relay and breaker,
(b) receivinf-end relay and breaker,
(c) both end relays,
(d) none of the above.

Q.219 A single transformer is protected by -
(a) an overcurrent relay,
(b) an earth-fault relay fed with residual voltage,
(c) both the above (a) and (b),
(d) none of the above.

Q.220 For transformers in parallel, a differential system of protection or intertripping system is essential to -

(a) trip the breakers on both sides of the transformer,
(b) keep a balance of power between input and output,
(c) fast clear the fault,
(d) none of the above.

Q.221 In intertripping protection of transformers, fault on any side provides tripping of breakers on either side through the use of -

(a) relays and pilot wires,
(b) direct interconnection of breakers of the two sides,
(c) none of the above.

Q.222 In any scheme for transformer protection, protection of transformer feeders -

(a) can also be included,
(b) can not be included,
(c) is certainly included,
(d) none of the above.

Q.223 In fault-throwing scheme of transformer protection, a fault of one type operates a relay which -

(a) trips the breaker on the two sides,
(b) creates an earth fault on the transformer forcing it to trip,
(c) provides an alarm indicating the fault,
(d) none of the above.

Q.224 In residual voltage tripping of a transformer or feeder, a residual voltage relay is used to trip the breaker, fed by transformer connected to the lines, secondary of which is connected in open delta feeding the relay coil. The tripping relay operates -

(a) due to unbalance in the three phases,
(b) under normal operation and keeps the breaker on,
(c) due to none of the above.

Q.225 The Buchholz relay protection of the transformer operates due to -

(a) turbulence or overheating of the transformer oil,
(b) overcurrent in the transformer winding,
(c) excessive voltage surge in the transformer winding,
(d) none of the above.

Q.226 The Buchholz relay is fitted -

(a) above the transformer oil tank,
(b) on the side of the oil tank dipped in the oil,
(c) in the pipe connecting the oil tank and the oil conservator,
(d) none of the above.

Q.227 Differential system for the protection of transformers or ganerators, senses the -
(a) input and output voltages,
(b) currents on the two sides of the windings,
(c) winding temperatures at different points,
(d) none of the above.

ABSTRACT

Electrical energy has been found to be the most convenient form of energy because of its transportability. Hence energy scene throughout the globe is dominated by electrical power generation and distribution. For reliable operations various switching, protection and control arrangements for the power systems have been developed.

5.1. GENERAL

Electrical power may be generated using following sources of energy :

(a) Fuels, such as coal, petroleums, natural gas;
(b) Water heads through flowing streams;
(c) Ocean tides and waves;
(d) Winds;
(e) Solar rays;
(f) Terrestrial heat;
(g) Atomic reactions.

Presently, coal and the water-heads are the main sources of electrical power, generated by large thermal and hydro power stations.

Power plant engineering is the art of selecting and placing the necessary power-generating equipment so that a maximum of return will result from a minimum of expenditure over the working life of a plant; and the operation of the completed plant in a manner to provide cheap, reliable and continuous power.

Generation of electrical power at the consumption point is feasible but due to economies of scale, most of the power is generated at the centralized large power stations and then transmitted to different load centres from where it is distributed to the load points. Thus a power system consists of generating plants, transmission system and the distribution system.

The variable load problem

The main problem of power plant engineering arises due to load variations as the different consumers load the power system at their will, differently at different times of the day, week and the year etc. In this connection following terms are defined :

Connected load : The sum of the continuous ratings of all the equipment and outlets on the customer's circuits.

Load factor : The average load for the specified time divided by the maximum load during that period.

Demand factor : This is referred to the customer and is his actual maximum demand divided by his connected load.

Diversity factor : Sum of the individual maximum demands of a subdivision of a power system, at actual values divided by the maximum simultaneous demand of the system.

Plant capacity factor : The actual energy production in a given period by a generating plant divided by the maximum possible energy that could be produced by the plant in the given period.

Plant use factor : The annual kWh production of a plant divided by the kilowatts of capacity times the number of hours, the plant was in operation.

Costing of electrical energy

Following elements enter into the cost of electrical energy to the consumer:

(a) Fixed element for the capital costs, interest and taxation, management, depreciation and general maintenance.

(b) Energy element for meeting the costs of fuels, labour, consumables, wastages and maintenance.

(c) Customer element proportional to the number of customers and includes the cost of secondary distribution, revenue collection, licensing fee, publicity and advertising etc.

(d) Investor's profit.

Factors and methods for billing

(a) Straight line meter rate.

(b) Block meter rate - rates vary for different blocks of power units consumed by the customer.

(c) Flat demand rate - the charges are fixed on the basis of the registered demand and independent of the actual consumption.

(d) Block Hopkinson demand rate - charges based on a maximum kw per month plus an energy block rate.

(e) Three-charge or 'Doherty' rate - A customer type charge is added to any of the methods of two-charges. Thus the billing charge is composed of customer or meter charge plus demand charge plus energy charge.

(f) Room rate charge - charges are fixed on the basis of the active rooms in the premises.

(g) Power factor charge - In addition to the above, customers may be encouraged to consume power at high power factors, by providing incentives or penalties based upon the power factor.

5.2. GENERATION OF ELECTRICITY

Fuels

(a) *Gas* : Natural gases is being made widely available through pipelines network. Its cost at the source is very low but at the consumption points increases due to pipelines and compressing costs. It is a very convenient fuel because of its being smokeless, ashless and easier controllability. Its main constituent is Methane with some quantities of ethane, nitrogen and carbon mono oxide. Its calorific value average around 1000 Btu per cu. ft.

(b) *Diesel oil* : Various grades of diesel oil are used as fuel in most of the captive power plants. Their calorific values vary between 17,500 Btu to 20,000 Btu per lb. For commercial power plants their costs are prohibitive.

(c) *Furnace oil* : Furnace oil is very often used to operate boilers and furnaces in industries but rarely this fuel is used for power generation.

(d) *Coal* : Thermal power plants provide more than 60 per cent of commercial power and they use coal as fuel. Various grades of coal are Anthracite, Bituminous, Subbituminous and Lignite. In general the coal is judged by its moisture, volatile matter and ash contents. The combustion of coal occurs when its atoms and some atoms of oxygen are joined through their electrons to form molecules. After combustion the number of electrons in the molecules remain same but with lower energy levels. This energy radiates from the molecules and impinges other molecules producing heat.

Thermal power plants

A thermal power plant consists of :

(a) *Boiler* : A device to generate heat through combustion of coal and transferring it to water to convert it into steam. The boiler may be one of the following types :

1. La Mont boiler
2. Loeffler boiler
3. Benson boiler
4. Velox boiler

(b) *Turbine* : Converts thermal energy of steam into mechanical energy of rotation.

(c) *Generator* : Converts mechanical energy into electrical energy.

The working cycle of a thermal power station is shown in Fig. 5.1.

For generation of electricity from flow of water, water is required to be at high altitudes, which has to be a natural occuring property. Hence choice of location of a power plant is limited by the natural water resources. The principal components of a hydro station are :

(a) *Water reservoir* : Water is stored in large quantity at a higher altitudo than the power station swo that water in available to the water wheel. From this reservoir, water is drawn through tunnels to the power station.

(b) *Hydraulic turbines* : Usually known as water-wheels have vanes around the periphery on which water under pressure hits to rotate it. The turbines are of two types, Reaction turbines and the Impulse turbines.

(c) *Generators* : The turbines drive the synchronous generators to generate three-phase electricity, which is the suitable transformed and transmitted to the distributing stations.

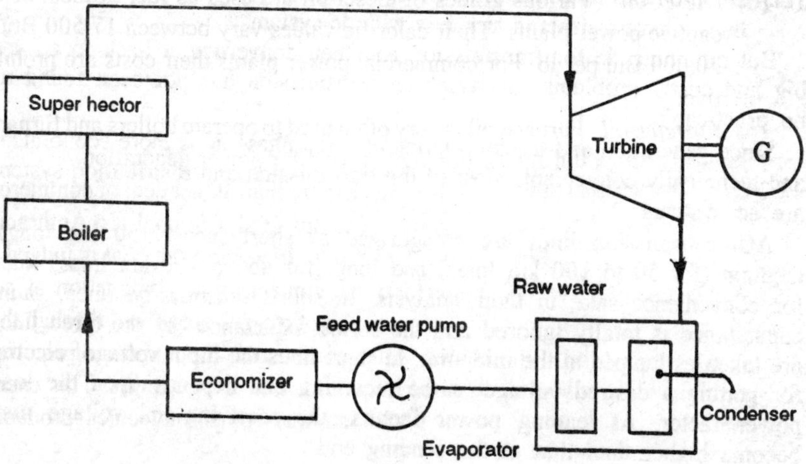

Fig. 5.1. Energy process cycle of a thermal power station.

Nuclear power stations

The main source of nuclear power is the fission of heavy elements like Uranium-235 and Thorium-232, which generate tremendous amount of heat. The generation of heat is moderated by the use of moderators such as graphite, berylium etc. The whole reaction is carried out in spherical shells known as reactors.

Heat from the reactors is taken away by coolants such as heavy water, organic liquids, mercury etc. which take the heat away to the utilization centres, where it is converted to electrical energy.

5.3. TRANSMISSION AND DISTRIBUTION

Power from the generating station is taken to the load centres or the distribution station through transmission system.

The transmission system consists of transformers for stepping-up, transmission lines, usually of overhead bare conductors, and the step-down transformers.

Transmission of power as dc has been found to be advantageous because of the following reasons :

(i) In dc there is no skin effect on the conductors hence lower size of conductors are possible.

(ii) AC transmission lines have power losses and the voltage drops due to line susceptance and the series inductance, which are absent in dc transmission.

(iii) DC produces static fields which cause no interference on communication lines. Corona effect is also less with dc system.

But obtaining dc for transmission and then converting it back to ac are big and costly problems for which dc transmission has not been found to be practicable.

Since generation and transformation of three-phase ac is more economical and technically convenient, most of the transmission and distribution system are ac systems.

AC transmission lines are catagorized as short (nearly 50 km long), medium (for 50 to 100 km lines) and long (for above 100 km lines) lines for convenience sake in their analysis. In short transmission lines, shunt capacitance is totally ignored and the series inductances of the three lines are taken as lumped at the midpoint. In short lines the input voltage required for getting a desired voltage at the receiving end depends upon the load power factor. At leading power factors, the receiving-end voltage may become higher than that at the sending end.

In medium length lines, the series inductance and shunt capacitances are taken as represented by Tee or Pie network, lumped at the middle of the line. For long lines, the series inductance and shunt capacitances are taken as uniformly distributed throughout the line, hence more accurate results are obtained.

For a transmission line, its characteristic impedance (Z) and propagation contant (Y) are given by :

$$Z = \sqrt{(z/y)} \quad \text{and} \quad Y = \sqrt{(zy)}$$

where z is the series impedance and y the shunt admittance per unit length of the line.

In generalized circuit constants form a line is represented by the relations:

$$V_s = AV_r + BI_r$$

and $I_s = C\,V_r + DI_r$

where V_s and V_r are sending-end and receiving-end voltage

I_s and I_r are sending-end and receiving-end currents

A, B, C, D are generalized circuit constants.

5.4. REPRESENTATION OF POWER SYSTEM

A power system is represented in per unit values of impedances with reference to some preselected base values of voltage and current. From these two values, other base values are obtained by the following relations :

Base kva = (base voltage in kv) x (base current in amps)

Base impedance = (base voltage in volts)/(base current in amps)

Base power in kw = base kva

Per-unit impedance of an element = (impedance in ohms)/base impedance

5.5. POWER SYSTEM OPERATION

(a) Load sharing

Incremental fuel cost of a unit at a given power output is the limit of the ratio of increase in cost of fuel input per hour to the corresponding increase in power output, as the increase in power output approaches zero.

For economic division of loads between the various units of a plant, all units must operate at the same incremental fuel cost.

(b) Sequence impedances

In unbalanced operations or fault currents, positive-, negative- and zero-sequence currents flow through the circuit. The voltage drops by the positive-sequence currents are due to the positive-sequence impedance of the circuit. Similarly, negative-sequence and zero-sequence volt drops are due to the flow of negative- and zero-sequence currents in the negative- and zero-sequence impedances of the circuit. Hence, we can say the positive-sequence impedance, negative-sequence impedance and zero-sequence impedance of circuit elements are different quantities and are valid for the respective sequence voltage and current parameters only. The different sequence parameters for a generator are related as follows :

$$V_{a1} = E_a - I_{a1}Z_1$$

$$V_{a2} = -I_{a2}Z_2$$

$$V_{a0} = -I_{a0}Z_0$$

where subscripts 1, 2 and 0 denote positive-, negative- and zero-sequence values.

(c) Power system stability

The tendency of a power system and its components to develop force as to maintain synchronism and equilibrium is the cause of stability of a power system.

The stability of a power system is defined as that attribute of the system, or part of the system, which enables it to develop restoring forces between the elements thereof, equal to or greater than the disturbing forces so as to restore a state of equilibrium between the elements.

The stability limit of a point of a power system, is the maximum power flow possible through that particular point in the system when the entire system or the part of the system is operating with stability.

The stability and the stability limit are different for steady-state and for transient state, which refer respectively to the situations of gradual changes and sudden disturbance of power flow in the system.

Any disturbance in the power flow produces a swing in the power angle of the generator, represented by the swing equation :

$$M \frac{d^2\delta}{dt^2} = P_a = P_s - P_e$$

where M is inertia constant of the machine in MJs/electrical degree. P_a is the accelerating power, P_s is the shaft power and P_e is negative of electric power developed.

The equal area criterion is used to study the stability of a machine in a power system.

6

Electrical Machines

Q.1 For an alternator, if frequency of oscillation is given by f = PN, P and N are respectively :

(a) No. of poles and speed in RPM,
(b) No. of phases and speed in RPS,
(c) No. of planes and No. of phases,
(d) None of above.

Q.2 In a d.c. machine, induced emf in the armature coil rotating in the magnetic field, will be at its maxima when the coil plane is -

(a) perpendicular to the magnetic field,
(b) parallel to the magnetic field,
(c) in any position,
(d) None of the above.

Q.3 In India the frequency of power supply being 50 Hz, theoritically the maximum rotating speed for all types of rotating machines is limited to -

(a) 3000 RPM, (b) infinity,
(c) 50 RPS, (d) None of the above.

Q.4 A.c. winding of rotating machines form closed path in itself. This statement is -

(a) wholly true, (b) partially true,
(c) wholly wrong, (d) nonsense.

Q.5 All d.c. machines armature windings have parallel paths equal to -

(a) two always, (b) one always,
(c) number of poles, always, (d) two or number of poles.

Q.6 A.C. machine armature windings can be of different types such as (i) lap, (ii) wave, (iii) concentric, (iv) eccentric.

(a) all the above are true,
(b) only (i) and (ii) are true,
(c) only (iii) and (iv) are true,
(d) only (iii) is true.

Q.7 In lap winding of a d.c. machine armature, commutator pitch is alwways-

(a) (+ or -) 1,
(b) m, where m is the multiplexity of the winding,
(c) (+ or -) 2,
(d) not fixed.

Q.8 D.C. machine armature winding configuration is -

(a) independent of number of poles,
(b) independent of speed of rotation,
(c) dependent upon number of poles and speed,
(d) dependent upon field current.

Q.9 In d.c. machines armature conducor size depends upon (i) KVA rating, (ii) voltage rating, (iii) current rating, (iv) conductor resistivity.

(a) all are true,
(b) only (ii), (iii) and (iv) are true,
(c) only (iii) and (iv) are true,
(d) only (iv) is true.

Q.10 Winding insulation class of an electrical machine is dependent upon -

(a) temperature rise in the machine,
(b) heat produced in the machins,
(c) thermal capacity of the machine,
(d) current rating of the machine.

Q.11 Electrical insulating materials are classified according to their

(a) dielectric strength, (b) temperature tolerence,
(c) current capacity, (d) thermal capacity.

Q.12 Single layer and double layer concept of windings is applicable to (i) d.c. machine armatures, (ii) d.c. field winding, (iii) a.c. armatures, (iv) transformers.

(a) all the above are true, (b) only (i) and (iii) are are true,
(c) only (i) is true, (d) only (iv) is true.

Q.13 For a particular d.c. machine, copper consumption in single-layer and double-layer windings will be in the ratio of nearly -

(a) 2:1, (b) 1:1,
(c) 1:2, (d) none above.

Q.14 M.M.F. produced by a winding distributed over the machine armature as compared to a concentrated winding is -

(a) slightly less, (b) slightly more,
(c) equal, (d) much more.

Q.15 M.M.F is to flux, is what E.M.F. is to current. Similarly the closest analogy is -

(a) voltage is to current,
(b) flux density is to flux,
(c) magnetic intensity is to flux density,
(d) current is to flux.

Q.16 The understood sinusoidality of the voltage waveform generated by an alternator is based upon (i) uniformity of flux distribution, (ii) sinusoidal distribution of flux, (iii) uniform rotary movement, (iv) sinusoidal rotation.

(a) only (i) and (iii) are true, (b) only (iii) is true,
(c) only (ii) is true, (d) only (iv) is false.

Q.17 Rotary machine armature cores are skewed to eliminate (i) tooth ripples, (ii) cogging, (iii) crawling, (iv) concentration of heat.

(a) all are true, (b) only (iv) is false,
(c) only (iii) and (iv) are false, (d) all are false.

Q.18 Armature reaction in d.c. machines can be said to be the -

(a) magnetic effect of curent flow in armature,
(b) opposition of armature against rotation,
(c) reactive power in armature,
(d) none of the above.

Q.19 In a transformer, the magnetizing flux produces (a) counter-emf in the primary, (b) working emf in the secondary, (c) iron losses in the core, (d) current in the secondary.

(a) all the above are true, (b) only first three are true,
(c) only first two are true, (d) only first is true.

Q.20 In a transformer the magnetizing flux is produced by

(a) primary and secondary MMFs jointly,
(b) residual magnetism of the core,
(c) primary emf only,
(d) primary voltage only.

Q.21 Like the counter-emf in the primary and the induced emf in the secondary of a transformer, such emfs are produced in (i) induction motors, (ii) synchronous motors, (iii) d.c. motors, (iv) autotransformers, also.

(a) all the above are true, (b) only first three are true,
(c) only first two are true, (d) only first one is true.

Q.22 Between two poles of a synchronous generator, the flux flow is regarded as sinusoidally varying, wrt.-

(a) magnetic axis,
(b) radius of the rotor,
(c) perpendicular to magnetic axis,
(d) none above.

Q.23 Speed of rotation of an armature-coil in a rotary machine is regarded to be sinusoidal wrt.-

(a) magnetic axis, (b) radius of rotor,
(c) periphery of the rotor, (d) none above.

Q.24 In a d.c. machine the rotational emf is given by e = p ϕ Z n /a, where, n is speed in rps and a is no. of parallel paths. (i) ϕ is flux density, (ii) Z is no. of coils, (iii) p is no. of poles.

(a) all (i), (ii) and (iii) are wrong,
(b) only (i) and (ii) are wrong,
(c) only (i) is wrong,
(d) all are true.

Q.25 The torque developed in any motor is considered to be uniform in the direction of -

(a) circumference of the rotor,
(b) radius of the rotor,
(c) perpendicular to the magnetic axis,
(d) magnetic axis.

Q.26 Speed of a d.c. motor is governed by the criteria - Back emf = applied voltage - armature resistive drop

(a) true,
(b) partially true,
(c) totally wrong,
(d) one more critoria is to be satisfied.

Q.27 The Lenz's left hand rule of electromagnetism is applicable to :

(a) D.C. Generator only, (b) D.C Motor only,
(c) A.C./D.C. Generators, (d) A.C./D.C. Motors.

Q.28 For a given type of 3-phase transformer, which of the following specifications will have the smallest size :

(a) 100 KVA, 11/0.4 KV, (b) 200 KVA, 22/0.8 KV,
(c) 100 KVA, 5.5/0.2 KV, (d) 200 KVA, 44/1.6 KV ?

Q.29 Which of the following laws apply to decide the direction of flux flow around a current-carrying conductor :

(a) Lenz's Law, (b) Kirchoff's law,
(c) Cork-screw rule, (d) none of the above.

Q.30 A synchronous generator with a given inductive load can be made to work at leading power factor by -

(a) Over-excitation,
(b) A synchronous motor in parallel with the generator,
(c) Over-sppeding,
(d) None.

Q.31 Which of the following statement(s) is/are correct ? (i) Direction of rotation of all electrical machines are fixed, (ii) All electrical machines can be rotated in any direction, (iii) direction of rotation of electrical machines depend upon the direction of curent, (iv) direction of elect. machine depend upon power factor.

(a) Only (i), (b) Only (i) and (ii),
(c) All, (d) None.

Q.32 Which type of the following cooling methods provide the best cooling ?

(a) ONWF, (b) Open type dry,
(c) OFWF, (d) ONAF.

Q.33 Which of the following excitation sources is technically the best for an alternator ?

(a) Full-wave rectified d.c., (b) D.C. generator,
(c) Three phase a.c., (d) Single phase a.c.

Q.34 For a single phase transformer of a given rating, the efficiency will be maximum :-

(a) For core type transformer,
(b) For shell type transformer,
(c) When copper and iron weights are equal,
(d) When copper and iron losses are equal.

Q.35 For a transformer (i) In core type only, iron weight is more than the copper weight, (ii) In shell type, copper weight is more than the iron weight, (iii) iron and copper losses are in the ratio of their respective weights,

(a) All the above are true, (b) None of above is true,
(c) Only (iii) is true, (d) only (i) and (ii) are true.

Q.36 Transformer size increases with increase in (i) volage rating, (ii) kva rating, (iii) frequency rating, (iv) power factor.
(a) only (ii) and (iii) are true. (b) only (i) and (iii) are wrong,
(c) Only (iii) and (iv) are true, (d) None is true.

Q.37 In a transformer (i) Flux and magnetizing current are always out of phase, (ii) On a leading load, secondary current can still be lagging, (iii) primary current can be leading, (iv) flux is always leading the primary back emf.
(a) All are true, (b) All are false,
(c) Only (ii) is false, (d) Only (i) and (iv) are true.

Q.38 In a unit ratio transformer if L_1 and L_2 are primary and secondary inductances and M, the mutual inductance between the two, ideally the following is true -
(a) $L_1 = L_2$, (b) $L_1 + L_2 + M = 0$,
(c) $L_1 + L_2 = M$, (d) None of the above.

Q.39 In a transformers, if N_1 and N_2 are primary and secondary turns and I_1 and I_2 are the primary and secondary currents, precisely, the equation $N_1 \times I_1 = N_2 \times I_2$
(a) Always true, (b) Never true,
(c) Rarely true, (d) Mostly true.

Q.40 In a transformer, if P is the load and k_1, k_2, k_3, k_4 are constants, transformer losses L are given by the equation :
(a) $L = k_1 \times P$ (b) $L = k_1 \times P^2 + k_2 \times P$
(c) $L = k_1 \times P^2 + k_2 \times P + k_3$ (d) None of the above.

Q.41 From the results of open-circuit and short-circuit tests on the transformer, we can get the following parameters of the transformer.
(a) Voltage regulation, (b) Efficiency,
(c) None of the above, (d) both (a) and (b).

Q.42 In a transformer, for primary and secondary terminals to be respectively H1, H2 and L1, L2, additive polarity means :
(a) H1 and L1 will be located closely (on the same end),
(b) H1 and L2 will be located closely,
(c) H1 is at higher potential than H2,
(d) None of the above.

Q.43 For three phase transformers, DY1 means
(a) Delta of yellow phase 1, (b) Difference of phase y is 1,
(c) Phase shift is 30 degrees, (d) None of the above.

Q.44 In 3 phase transformers, V connection means :
(a) Two legs of star connection,
(b) Transformer voltage is variable,
(c) Vertical arrangement of terminals,
(d) None of the above.

Q.45 For parallel operation of two 3-phase transformers, both transformers must be of (i) Same voltage ratio, (ii) Identical polarity, (iii) same phase shift, (iv) same voltage rating.
(a) All the above are sufficient,
(b) All the above are not essential,
(c) Any two of the above are enough,
(d) One more essential condition is missing.

Q.46 Transformer T-connection is used for (i) Tertiary winding, (ii) 2-to-3 phase conversion, (iii) inverse of (ii), (iv) Tetra phasing.
(a) All the above are false, (b) Only (ii) and (iii) are true,
(c) Only (iii) is true, (d) Only (iv) is true.

Q.47 In transformers, the core and shell type refer to (i) 1-phase transformers, (ii) 3-phase transformers, (iii) only aircooled transformers.
(a) Only (i) is true, (b) Only (iii) is true,
(c) Only (i) and (ii) are true, (d) None is true.

Q.48 Which one is true -
(a) A variac is an autotransformer,
(b) an autotransformer is a variac,
(c) variac and autotransformer are synonymous,
(d) variac and autotransformer are unrelated.

Q.49 An induction regulator is (i) a frequency changer, (ii) a voltage variator and phase-shifter, (iii) a power factor controller, (iv) an induction motor outwardly modified..
(a) only (i) and (ii) are true, (b) only (ii) and (iv) are true,
(c) only (ii) is true, (d) only (iii) is true.

Q.50 An induction motor is also known as universal transformer because it can provide (i) variable voltage, (ii) variable frequency, (iii) variable current.
(a) only (i) and (ii) are true, (b) only (i) and (iii) are true,
(c) all are true, (d) none is true.

Q.51 A transformer's on-load tap-changer is so called because (i) its sparking on load is not damaging, (ii) it does not interrupt load current, (iii) it is very heavy, (iv) it can be operated with full load.
(a) only (iii) is true. (b) only (i) and (ii) are true,

(c) only (i),(ii) and (iv) are true, (d) all are true.

Q.52 For a current transformer (i) secondary has no voltage but only the current, (ii) current is independent of line voltage, (iii) has no primary, (iv) is a current operated device.
(a) only (iii) is true, (b) only (i) and (ii) are true,
(c) only (ii) and (iii) are true, (d) Only (ii) and (iv) are true.

Q.53 Heat generated in a transformer is depndent upon (i) load on the transformer, (ii) Voltage of the transformer, (iii) current of the transformer.
(a) all are true, (b) only (i) and (ii) are true,
(c) only (i) is true, (d) none is true.

Q.54 On a sudden short-circuit fault on the line fed by a transformer, the mechanical stresses are maximum at the transformer :
(a) core, (b) core yoke,
(c) coils, (d) coil ends.

Q.55 In electromagnetic equipment, energy transfer takes place (i) conductively, (ii) inductively, (iii) convectively, (iv) radiatively.
(a) all the above are true, (b) only first three are true,
(c) only first two are true, (d) only first one is true.

Q.56 The armature reaction is fruitfully utilized in -
(a) synchronous machines, (b) induction machines,
(c) d.c.machines, (d) metadynes.

Q.57 Compensating windings in synchronous machines are provided to overcome the effect of -
(a) armature resistive drop, (b) magnetic flux leakage,
(c) armature reaction, (d) core iron losses.

Q.58 Theoretically, the amature winding of an induction motor can be placed on the rotor, with the rotor winding on the stator.
(a) it is not true, (b) it is partially true,
(c) it is true, (d) it makes no sense.

Q.59 Squirrel cage induction motor is so called for its (i) armature construction, (ii) rotor winding construction, (iii) simplicity, (iv) robustness.
(a) all the above are true, (b) only last three are true,
(c) only (ii) is true, (d) only (i) is true.

Q.60 In induction motors, major transmission of electrical power is -
(a) conductive, (b) inductive,
(c) magnetic, (d) none above.

Q.61 In induction motors, stator and rotor currents have the relationship -
(a) rotor current > stator current,
(b) rotor current < stator current,
(c) rotor current and stator current have no fixed relationship,
(d) none of the above.

Q.62 In induction motors, the outer body of the stator provides (a) mechanical support, (b) path for magnetic flux, (c) path for induced currents, (d) path for magnetic field.
(a) all the aboveare true, (b) only first three are true,
(c) only first two are true, (d) only first one is true.

Q.63 As compared to d.c. and synchonous machins, induction machines airgap is -
(a) comparable, (b) very small,
(c) very large, (d) same.

Q.64 For induction motors, axial lengths of the stator and rotor cores should be -
(a) equal, (b) equal and aligned,
(c) unequal, (d) equal but not aligned.

Q.65 In induction motors, magnetic flux passes through the shaft also.
(a) true, (b) may be true,
(c) not true.

Q.66 Small induction motor core should be of-
(a) silicon steel, (b) grain-oriented silicon steel,
(c) ordinary iron, (d) special En steel.

Q.67 In induction motors, stator and rotor windings could be identical.
(a) false. (b) true.
(c) only partially.

Q.68 For production of electromagnetic torque, interaction between a magnetic field and a current is essential. In induction motors, the provisions are -
(a) field by stator and the current by rotor.
(b) field by rotor and the currentby stator,
(c) there is no field,
(d) the two roles can be treated either way.

Q.69 The rotating magnetic field in a 3-phase induction motor is produced by -
(a) phase shift of the three phases,
(b) each phase independently,

(c) the distributiveness of a phase winding,
(d) none of the above.

Q.70 In induction motors, the term 'synchronous speed' is -
(a) rotor speed, (b) rotating field speed,
(c) irrelevent, (d) fictitious.

Q.71 The speed of an induction motor can be said to be _
(a) hyper-synchronous, (b) hypo-synchronous,
(c) super-synchonous, (d) none of above.

Q.72 An induction motor can attain synchronous speed only when -
(a) load on the motor is zero,
(b) motor power factor is unity,
(c) motor supply frequency is zero,
(d) none of the above.

Q.73 Synchronous speed of an induction motor is (i) proportional to the supply frequency, (ii) inverse of (i), (iii) proportional to the number of poles, (iv) inverse of (iii).
(a) only (i) is true. (b) question is irrelevent,
(c) only (i) and (iii) are true, (d) only (i) and (iv) are true.

Q.74 Eddy current losses in an induction machine can be reduced by -
(a) reducing lamination thickness,
(b) reducing number ofturns in stator winding,
(c) increasing winding conductor area.
(d) none of the above.

Q.75 In induction motors, rotor and stator lamination thicknesses t_1 and t_2 respectively, should be related as -
(a) $t_1 > t_2$, (b) $t_1 < t_2$,
(c) $t_1 = t_2$, (d) it is irrelevent.

Q.76 In a slipring induction motor, the speed of rotation can be reversed by (i) interchanging any two stator supply connections, (ii) by interchanging any two slipring connections,
(a) by (i) only (b) by (i) or (ii),
(c) by (i) and (ii) both, (d) by (ii) only.

Q.77 In a slipring induction motor, the roles of stator and rotor windings are reversible, except for the voltage ratings.
(a) true, (b) false,
(c) partially true, (d) nonsense.

Q.78 Induction motor running torque can be increased by -
(a) increasing stator impedance, (b) increasing rotor resistance,
(c) by reducing rotor resistance, (d) none of the above.

Q.79 Induction motor starting torque can be increased by -

(a) increasing stator resistance, (b) increasing rotor resistance,
(c) decreasing rotor resistance, (d) none of the above.

Q.80 Induction motor running torque is approximately -

(a) proportional to its slip,
(b) inversely proportional to slip,
(c) proportional to the square of slip.
(d) none of the above.

Q.81 If an induction motor is driven at synchronous speed, its rotor voltage will be (i) zero, (ii) of zero frequency, (iii) in phase with stator voltage.

(a) only (i) is true, (b) only first two are true,
(c) all are true, (d) all are wrong.

Q.82 In an induction motor starting torque is always -

(a) less than its maximum torque,
(b) equal to its maximum torque,
(c) dependent upon the rotor resistance,
(d) independent of maximum torque.

Q.83 Induction motor torque is maximum when -

(a) rotor resistance and reactance are zero,
(b) rotor resistance and reactance are equal,
(c) slip is zero,
(d) rotor current is maximum.

Q.84 If an induction motor has rotor winding, then only it can be used as (i) a drive with high starting torque, (ii) an induction regulator, (iii) a frequency changer, (iv) an induction generator.

(a) only (i) is true, (b) only (i) and (ii) are true,
(c) all, except (iv) are true, (d) all are true.

Q.85 A 3 percent change in the supply voltage of an induction motor will change its torque nearly by -

(a) 3 percent, (b) 6 percent,
(c) 9 percent, (d) zero percent.

Q.86 Slip of an induction motor can be obtained by the following parameters of motor stator and rotor -

(a) frequencies, (b) voltages,
(c) currents, (d) resistances.

Q.87 The ratio of rotor and stator voltages of an induction motor is proportional to (i) speed, (ii) inverse of speed, (iii) slip, (iv) inverse of slip.

(a) only (i) and (iii) are true, (b) only (ii) and (iii) are true,
(c) only (i) and (iii) are true, (d) only (iii) is true,

Q.88 Double cage induction motor provides -
(a) high starting torque,
(b) high starting and running torque,
(c) low starting current,
(d) more robustness.

Q.89 In double cage induction motors, outer and inner cage resisances are respectively -
(a) high and low, (b) low and high,
(c) equal, (d) uncomparable.

Q.90 In induction motors, number of slots in stator and rotor -
(a) should have an integral ratio,
(b) should not have certain combinations,
(c) are independently choosen,
(d) should be equal.

Q.91 In an induction motor, for maximum torque at the starting, rotor resistance should be nearly equal to -
(a) rotor reactance, (b) total leakage reactance,
(c) stator resistance, (d) none of the above.

Q.92 In an induction motor, if the load torque exceeds the maximum torque of the motor, the motor speed will fall to -
(a) zero,
(b) nearly 10 percent of normal speed,
(c) crwling speed,
(d) none of the above.

Q.93 The circle diagram of an induction motor represents the locus of -
(a) primary current, (b) motor torque,
(c) output power, (d) none of the above.

Q.94 Induction motor iron, friction and windage losses can be roughly obtained by light running the motor and measuring the input power.
(a) true, (b) not true,
(c) partially true,

Q.95 Induction motor iron losses are -
(a) almost constant, (b) vary in proportion with load,
(c) depend upon stator current, (d) vary with rotor speed.

Q.96 What open circuit test is for a transformer, light running is for induction motor.

(a) true, (b) not true,

(c) partially true,

Q.97 What short circuit test is for a transformer, for induction motor, the equivalent test is -

(a) light running test,

(b) blocked rotor test,

(c) running the motor with rotor short-circuited.

(d) none of the above.

Q.98 For a plain induction motor, if X_1 is the stator reactance and X_{2e} the motor rotor equivalent reactance, it can be safely assumed, that -

(a) $X_1 = X_{2e}$, (b) $X_1 = .5 \times X_{2e}$,

(c) $X_1 = 2 \times X_{2e}$, (d) none of the above.

Q.99 For special high starting torgue motors, rotor reactance is usually

(a) more than the stator reactance,

(b) equal to the stator reactance,

(c) less than the stator reactance,

(d) none of the above.

Q.100 Some induction motors can be run at (i) slightly lower than the synchronous speed, (ii) at 50 percent of the usual speed, (iii) at any speed.

(a) all are true, (b) only first two are true,

(c) only first is true, (d) none is true.

Q.101 Cross field theory is an alternative theory for -

(a) rotating field theory of a.c. machines,

(b) explaining the working of amplidynes,

(c) explaining quadrature axis field for d.c. machines,

(d) none of the above.

Q.102 The concept of rotor equivalent values for induction motors, arises due to -

(a) turns ratio of stator and rotor windings,

(b) number of poles of stator and rotor windings,

(c) stator and rotor resistance values,

(d) stator and rotor current values.

Q.103 For a transformer, the concept of secondary equivalent values arise due to the difference in primary and secondary -

(a) turns, (b) coil sizes,

(c) conductor sizes, (d) leakage fluxes.

Q.104 In electrical machines, the effective resistances of windings are greater than the ohmic values, due to -

(a) skin effect, (b) proximity effect,
(c) bends in the conductors, (d) any other factor.

Q.105 Crawling in induction motors is caused by -

(a) pulsations in air gap flux,
(b) sinusoidal variation of input current,
(c) overloading of the motor,
(d) any other reason.

Q.106 Parasitic torque in induction motors, is the torque produced by -

(a) airgap flux pulsations, (b) input supply fluctuations,
(c) load fluctuations, (d) any other reason.

Q.107 In induction motors, airgap flux pulsations are suppressed by -

(a) skewing,
(b) capacitors in parallel with the motor,
(c) double-caging the rotor,
(d) any other method.

Q.108 In induction motors, skewing means -

(a) creation of slant in the core slot,
(b) tapering the rotor,
(c) keeping the rotor slightly unaligned with the stator,
(d) any other thing.

Q.109 Induction motors are started with reduced input voltage to -

(a) reduce starting current, (b) speedup gradually,
(c) reduce starting torque, (d) eliminate crawling.

Q.110 Star-delta starter is used to start only squirrel-cage induction motors.

(a) true, (b) false,
(c) partially true,

Q.111 For starting slipring induction motors, external resistance is introduced in the rotor circuit. it (i) increases starting current, (ii) increases starting torque, (iii) decreases starting current, (iv) decreases starting torque.

(a) only (i) is true, (b) only (i) and (ii) are true,
(c) only (ii) and (iii) are true, (d) only (ii) is true.

Q.112 In double-cage induction motors, the two cages work -

(a) in series with each other, (b) in parallel with each other,
(c) one at starting time and the other at running time.

Q.113 In squirel-cage induction motors, speed control can be achieved by (i) stator voltage control, (ii) pole changing, (iii) stator frequency control, (iii) stator reactance control.

(a) all are true, (b) only first three are true,
(c) only first two are true, (d) only first one is true.

Q.114 In slipring induction motors, speed control is normally done by (i) rotor resistance control, (ii) stator frequency control, (iii) rotor voltage control, (iv) pole changing.

(a) all are true, (b) only first three are true,
(c) only first two are true, (d) only first one is true.

Q.115 Schrage motor is basically an induction motor with provision for speed control.

(a) true, (b) partially true,
(c) false.

Q.116 Position selsyn is a device which (i) indicates position, (ii) has two complimenary units, (iii) operates wih a.c. and d.c., (iv) can be used to indicate difference of the two positions.

(a) all the above are true, (b) only first three are true,
(c) only first two are true, (d) only first one is true.

Q.117 A differential selsyn has

(a) three identical units,
(b) indicates difference of the two angular positions,
(c) rotors of the three units are fed with a.c.,
(d) stators are fed with a.c..

Q.118 Any induction motor can be used as induction generator provided (i) stator is connected to the a.c. supply, (ii) stator is parallelled with capacitors, (iii) rotor is of wound type, (iv) rotor is driven in opposite direction.

(a) only (i) is true, (b) only (i) and (ii) are true,
(c) only (iv) is false, (d) all are true.

Q.119 An induction generator, connected to a power bus, its frequency is decided by -

(a) the frequency of power it is connected to,
(b) its rotation speed,
(c) resonance of stator circuit,
(d) number of poles.

Q.120 An induction generator must always be run at -

(a) synchronous speed, (b) super-synchronous speed,
(c) sub-synchronous speed, (d) reverse speed.

Q.121 Characteristics of an induction generator action are -
(a) duplicate of motor action characteristics,
(b) mirror image of motor action characteristics,
(c) inverted of motor action characteristics,
(d) independent of motor action characteristics.

Q.122 An induction generator cannot be operated in parallel with a synchronous generator.
(a) true,
(b) partially true,
(c) false.

Q.123 In an induction generator, the generator power factor is decided by -
(a) load power factor,
(b) generator characteristics only,
(c) load and generator both,
(d) none ofthe above.

Q.124 An induction generator can be run with loads of -
(a) any power factor,
(b) high and lagging power factor,
(c) leading power factor,
(d) unity power factor.

Q.125 Induction generators main limitation is their inherent -
(a) low power factor,
(b) high leading power factor,
(c) unity power factor,

Q.126 In single phase induction motors thre is no rotating magnetic field-
(a) true,
(b) false,
(c) nonsense.

Q.127 Single phase induction motor stator produces two torques working in opposite directions.
(a) true,
(b) false,
(c) nonsense.

Q.128 Single phase induction motors are used in (i) ceiling fans, (ii) some domestic drives, (iii) portable machine tools, (iv) cassette players.
(a) only (i) is true,
(b) only first two are true,
(c) only (iv) is false,
(d) all are true.

Q.129 Single phase induction motors have a capacitor to make it self-starting.
(a) true,
(b) false,
(c) partially true.

Q.130 Various methods to create starting torque in a single-phase inductin motor are (i) starting winding only, (ii) pole shading, (iii) starting winding with a capacitor, (iv) starting winding with a resistor.

(a) all are true, (b) only first three are true,
(c) only first two are true, (d) only first oneis true.

Q.131 Capacitor-start single phase induction motors have a centrifugal switch to disconnect the starting winding.

(a) true, (b) false,
(c) partially true.

Q.132 Single phase induction motors are normally of -

(a) low power factor, (b) unity power factor,
(c) high power factor,
(d) leading power factor due to the presence of capacitor.

Q.133 In comparison to water-wheel alternators, turbo-alternators have (i) high length/diameter ratio, (ii) high speed, (iii) more number of poles, (iv) more saliency in poles.

(a) only (i) is true, (b) only first two are true,
(c) only last one is false, (d) all are true.

Q.134 H class of insulating materials can work upto a temperature of -

(a) 130 degree celcius, (b) 150 degree celcius,
(c) 180 degree celcius, (d) 210 degree celcius.

Q.135 'A' class insulation materials have a limit of working temerature as -

(a) 90 degree celcius, (b) 110 degree celcius,
(c) 130 degree celcius, (d) 150 degree celcius.

Q.136 'B' class insulation materials have a limit of working temerature as -

(a) 90 degree celcius, (b) 110 degree celcius,
(c) 130 degree celcius, (d) 150 degree celcius.

Q.137 Voltage regulation of generators and transformers is a measure of -

(a) voltage drop due to loading,
(b) voltage fluctuations due to speed fluctuations,
(c) voltage amplitude,
(d) frequency response.

Q.138 In electrical machines 'saliency' means (i) outward projection, (ii) silent operation, (iii) selling potential, (iv) continuous operation.

(a) only (i) is true, (b) only first two are true,
(c) only (iv) is false, (d) none is true.

Q.139 Open circuit characteristic of an alternator is also known as (i) no-load characteristic, (ii) magnetic characteristic, (iii) saturation characteristic, (iv) external characteristic.

(a) all the above are true, (b) only first three are true,
(c) only firsttwo are true, (d) only firstone is true.

Q.140 Open circuit characteristic of an alternator droops due to -

(a) resistive drop in the armature,
(b) magnetic saturation,
(c) field resistive drop,
(d) any other reason.

Q.141 Short circuit characteristic of an alternator is -

(a) a straight line, (b) an arch,
(c) a combination of (a) and (b), (d) any other.

Q.142 Synchronous impedance of an alternator is given by Z = E/I, where E and I are respectively -

(a) Load voltage and load current,
(b) Open circuit voltage and current,
(c) Short ciruit voltage and current,
(d) Open-circuit voltage and short-circuit current,

Q.143 Slip test is a test applicable to (i) induction machines, (ii) synchronous machines, (iii) d.c. machines.

(a) only (i) is true, (b) only (i) and (ii) are true.
(c) only (ii) is true, (d) all are true.

Q.144 Direct-axis and quadrature-axis reactances of an alternator are caused by (i) magnetic pole saliency, (ii) distribution of the winding, (iii) alternation of a.c. wave, (iv) magnetic saturation.

(a) only (i) is true. (b) only first two are true,
(c) only last one is true. (d) all are false.

Q.145 Short circuit ratio of an alternator is the ratio of -

(a) Open circuit voltage and short circuit current,
(b) Short circuit voltage and current,
(c) Field current ratio for rated o.c. voltage and s.c. current.
(d) Synchonous reactance and actual reactance.

Q.146 In a synchronous motor, the poles of the field produced by armature and those of the field are rigidly locked, producing synchronous speed of the motor.

(a) true, (b) false,
(c) nonsense,

Q.147 For a synchronous motor, if ohmic losses are ignored, total torque developed is proportional to -

(a) electrical power input, (b) power factor,
(c) power angle, (d) magnetic saturation.

Q.148 In a synchronous motor, the exact cause of torque is the product of counter emfs and the component of currents in phase therewith.

(a) true, (b) partially true,
(c) false, (d) nonsense.

Q.149 If a synchonous generator, connected to a bus, is momentarily disconnected from the supply and reconnected,

(a) it will regain generation,
(b) it will start behaving like a motor,

Q.150 For synchronous machine, transition from generator action to motor action, can achieved by -

(a) reversal of its field,
(b) reduction of its field,
(c) reversal of armature connections,
(d) reversing the direction of rotation.

Q.151 For a synchronous motor, running on a constant voltage bus, if the excitation is gradually increased,

(a) the motor will start behaving like a generator,
(b) motor's power factor will improve,
(c) motor's speed will change,
(d) nothing notable will happen.

Q.152 A synchronous condenser is -

(a) an under-excited synchronous motor,
(b) an over-excited synchronous motor,
(c) a synchronous generator with super-synchronous speed.
(d) a resonating condenser circuit.

Q.153 For a synchronous motor, normal excitation means -

(a) excitation to provide normal speed,
(b) excitation just sufficient to take up normal load,
(c) excitation for unity power factor,
(d) full load current of the exciter.

Q.154 For electrical machines, V-curves mean characteristics of -

(a) synchronous generator, (b) synchronous motor,
(c) induction motor, (d) d.c. generator.

Q.155 For a synchonous motor, if excitation is gradually increased, its armature current -

(a) first increases and then falls,
(b) steadily decreased,
(c) first decreases and then increases,
(d) remains steady.

Q.156 Hunting in a synchronous motor means -
(a) pulsation of rotor speed of rotation,
(b) tooth ripples in airgap,
(c) armature current pulsation due to varying iron losses.
(d) deflection in rotor shaft.

Q.157 Hunting in synchronous motors may be caused by (i) voltage fluctuations, (ii) load fluctuations, (iii) excitation fluctuations, (iv) eddy current.
(a) all are true, (b) only (iv) is false,
(c) only first two are true, (d) only first one is true.

Q.158 Hunting in synchronous motors is suppressed by (i) eddy currents in the core, (ii) damper windings on pole faces, (iii) hysterisis losses, (iv) excitation hysterisis.
(a) only first one is true,
(b) only first two are true,
(c) only first three are true,
(d) none of the above (a),(b) and (c) is true.

Q.159 In synchronous motors, damper windings carry current (i) always, (ii) during hunting, (iii) during starting.
(a) all are true, (b) only (i) is false.
(c) only (i) is true. (d) only (iii) is true.

Q.160 A plain synchronous motor is not self-starting because (i) it develops torque at synchronous speed only, (ii) the starting current is too low to produce sufficient torque, (iii) field excitation produces braking torque,
(i) all are true, (b) only first two are true,
(c) only first one is true. (d) none is true.

Q.161 In synchronous motors, starting is aided by (i) pole saliency, (ii) damper winding, (iii) eddy currents, (iv) magnetic hysterisis.
(a) all are true, (b) only first three are true,
(c) only first two aretrue, (d) only first one is true.

Q.162 For parallel operation of two alternators, the two should have (i) same speed, (ii) same voltage, (iii) same phase sequence, (iv) same direction of rotation.
(a) all the above are essential, (b) all the above are enough,
(c) all the above are neither essential nor enough.

Q.163 In alternators, synchronism means (i) equality of voltage, (ii) equality of frequency, (iii) equality of phase sequence, (iv) equality of speeds.

(a) the above is true.

(b) the above conditions are neither enough nor essential,

(c) the above conditions are sufficient.

Q.164 Synchroscope is an instrument used for -

(a) measuring synchronous speed,

(b) verifying synchronizing conditions,

(c) synchronizing alternators,

(d) synchronizing induction generators.

Q.165 Synchronizing power of an alternator is its inherent power to -

(a) pickup to original speed,

(b) synchronize with the other machine in parallel.

(c) nonsense.

Q.166 For two alternators running in parallel if one lags behind, it starts behaving like -

(a) a motor,

(b) an alternator with more synchronizing power,

(c) an induction motor,

(d) an over-excited alternator.

Q.167 For two alternators running in parallel, if one leads the other, the leader starts behaving like -

(a) a motor, (b) an alternator with more load,

(c) an induction generator, (d) unaffected.

Q.168 The main reason and cause of synchronizing power of an alternator is its reactance.

(a) true, (b) false,

(c) nonsense.

Q.169 The loading of parallel operating transformers is dependent upon their -

(a) internal impedances, (b) magnetizing currents,

(c) phase shift, (d) power inputs.

Q.170 The loading of parallel operating alternators is dependent upon their -

(a) internal impedances, (b) power inputs,

(c) excitations (d) any other factor.

Q.171 The speed of a d.c. series motor at no-load is considered to be infinity, because -

(a) of weak field, (b) high armature voltage,

(c) wrong notion, (d) low armature current.

Q.172 If a d.c. shunt motor is reconnected and run as a series motor with no-load, it (i) may fail to rotate, (ii) its speed will be very low, (iii) its field will be of almost normal value, (iv) its speed will be very high.

(a) only (i) is true, (b) only first two are true,
(c) only (iv) is false, (d) only (iv) is true.

Q.173 Speed of a series motor, at fixed load, can be increased by (i) increasing input voltage , (ii) putting a divertor across field, (iii) putting a divertor across armature.

(a) all the above are true, (b) only first two are true.
(c) only last two are true, (d) none is true.

Q.174 If an alternating voltage of rated value is applied to a d.c. series motor, it will -

(a) run at very small speed,
(b) it will get damaged,
(c) it will run almost as usual,
(d) it will run at synchronous speed.

Q.175 The direction of rotation of a d.c. series motor can be reversed by (i) reversing the input supply, (ii) reversing the field terminals only, (iii) reversing the armature terminals only.

(a) all the above are true, (b) only (i) is false.
(c) only (ii) is false, (d) only (iii) is false.

Q.176 Skin effect in conductors will be low if (i) area of cross-section is low, (ii) frequency of supply is low, (iii) conductivity is high, (iv) conductor material is pure.

(a) all the above are true, (b) only first three are true,
(c) only first two are true, (d) only first one is true.

Q.177 What will happen if the field coil of a d.c. series motor is open-cicuited -

(a) motor will draw heavy current and stop.
(b) motor will stop and fail to restart,
(c) motor will pickup very high speed,

Q.178 D.C. series motors are used for -

(a) high starting torque applications,
(b) low starting current,
(c) constant speed operation,
(d) constant torque.

.179 D.C. series motor, if started at no-load, it will -

(a) run smoothly, (b) pickup very high speed,

(c) fail to start, (d) draw very heavy current.

Q.180 In d.c.motors, interpole windings are connected -
(a) in parallel with the armature, (b) in series with the armature,
(c) in series with the field, (d) in parallel with the field.

Q.181 In d.c. machines, interpoles are provided for -
(a) stablizing speed at varying loads.
(b) compensating for armature reaction,
(c) field compounding,
(d) reducing starting current.

Q.182 In d.c.machines, armature reaction is -
(a) magnetisation due to armature conductor currents,
(b) armature voltage drop,
(c) armature's tendency to oppose rotation,
(d) armature current fluctuation.

Q.183 In d.c. machines, armature reaction is minimized by (i) compensating winding at pole faces, (ii) interpoles, (iii) shifting of carbon brushes, (iv) increasing airgap.
(a) all are true, (b) only last three are true,
(c) only last two are true, (d) only last one is true.

Q.184 In d.c. machines, armature reaction causes (i) distortion of the magnetic field, (ii) sparking at carbon brushes, (iii) the production of a cross-magnetization field, (iv) field weakening.
(a) all are true, (b) only first three are true,
(c) only first two are true, (d) only first one is true.

Q.185 In d.c. motors, armature volt drop causes (i) drop in speed, (ii) lowering of efficiency, (iii) field weakening, (iv) sparking at carbon brushes.
(a) all are true, (b) only first three are true,
(c) only first two are true, (d) only first one is true.

Q.186 In d.c. motors, armature current depends upon (i) field intensity, (ii) speed of rotation, (iii) armature resistance, (iv) load.
(a) all are true, (b) only first two are true,
(c) only first three are true, (d) only last two are true.

Q.187 In d.c. machines, armature resistance depends upon (i) conductor area, (ii) type of winding - lap or wave, (iii) number of turns per coil, (iv) armature reaction.
(a) all are true, (b) only first three are true,
(c) only first two are true, (d) only first one is true.

Q.188 In d.c.motors, armature resistance is the main cause of power loss.

(a) true, (b) false,
(c) nonsense.

Q.189 In d.c. shunt motors, field circuit should be put on first, then only armature should be energized.

(a) true, (b) false,
(c) nonsense, (d) partially true.

Q.190 In regular reversing of d.c. shunt motors, (i) armature reversal should be adopted, (ii) field reversal should be adopted.

(a) only (i) is true, (b) only (ii) is true,
(c) both are true.

Q.191 In a running and loaded d.c. shunt motor, if field circuit is accidently opened -

(a) motor will calmly come to standstil,
(b) motor will draw a very heavy current and may get damaged,
(c) motor will overspeed.

Q.192 When loaded, d.c. shunt motor's speed -

(a) lowers, (b) remains same,
(c) does not change.

Q.193 To get a steady speed from d.c. motors, (i) cumulative compounding, (ii) differential compounding, (iii) use of interpoles, is adopted.

(a) only (i) is true, (b) only (i) and (iii) are true,
(c) only (ii) is true, (d) only (iii) is true.

Q.194 D.C shunt motor's direction of rotation can be changed by (i) reversing power supply, (ii) reversing armature connections, (iii) reversing field connections.

(a) only (i) is true, (b) only (i) and (ii) are true,
(c) only (ii) and (iii) are true, (d) all are true.

Q.195 If a d.c. shunt motor is supplied with a very low frequency a.c. voltage, it will -

(a) not run at all,
(b) run with carbon brush sparking,
(c) oscillate with sparks at carbon brushes.

Q.196 Speed of a d.c. shunt motor can be increased by (i) increasing armature voltage, (ii) incresing field current, (iii) reducing armature voltage, (iv) reducing field current.

(a) only first two are true,
(b) only (i) and (iv) are true,
(c) only (ii) and (iii) are true,
(d) only last two are true.

Q.197 In a d.c. shunt motor, if field is varied, motor's (i) torque, (ii) horse power, also vary.
(a) only (i) is true, (b) only (ii) is true,
(c) both are true, (d) both are false.

Q.198 In a d.c. shunt motor, if armature voltage is varied, motor's (i) torque, (ii) horse power, also vary.
(a) only (i) is true, (b) only (ii) is true,
(c) both are true, (d) both are false.

Q.199 To run a d.c. motor with constant horse power, its (i) armature voltage, (ii) field current, must remain constant.
(a) only (i) is true, (b) only (ii) is true,
(c) both are true, (d) both are false.

Q.200 To run a d.c. motor with constant torque, its (i) armature voltage, (ii) field current, must be steady.
(a) only (i) is true, (b) only (ii) is true,
(c) both are true, (d) both are false.

Q.201 In a d.c. compound motor, (i) series field, (ii) shunt field, (iii) interpole, (iv) equalizing ring, must be present.
(a) only (i) is true, (b) only (ii) is true,
(c) only first three are true, (d) only first two are true.

Q.202 Differential compounding is not used in d.c. motors.
(a) true, (b) false,
(c) partially true, (d) nonsense.

Q.203 In d.c. compound motors, series and shunt field windings are wound on -
(a) alternate main poles, (b) all main poles,
(c) all, including interpoles, poles.

Q.204 In d.c. compound motors, usually, the series field ampere-turns and the shunt field ampere-turns have the following relationship -
(a) series > shunt, (b) shunt > series,
(c) Shunt = series,

Q.205 The series field winding in a d.c. compound motor, (i) increases torque, (ii) increases speed.
(a) only (i) is true (b) only (ii) is true,
(c) both are true.

Q.206 In d.c. shunt generators, on loading, the output voltage dips due to -
(a) armature resistance, (b) field weakening,
(c) overheating.

Q.207 D.C. series generators are rarely used.
(a) true, (b) false,

Q.208 In d.c. series motors, dynamic braking is achieved by -
(a) running the motor as a generator,
(b) overloading the motor,
(c) shorting the armature.

Q.209 In motor operation, plugging means -
(a) providing supply to reverse the motor,
(b) blocking the rotor,
(c) shorting the armature.

Q.210 In electric motors, commutator is used to -
(a) convert a.c. to d.c., (b) convert d.c. to a.c.,
(c) both of (a) and (b), (d) none of (a) and (b).

Q.211 For a self-excited d.c. shunt generator, one of the essential conditions of generation is that some residual magnetism in the generator core should be available.
(a) true, (b) partially true.
(c) false.

Q.212 For a self-excited d.c. shunt generator, its field circuit resistance should be -
(a) equal to the critical resistance,
(b) more than the critical resistance,
(c) less than the critical resistance,
(d) equal to the armature resistance.

Q.213 In unsaturated region, reluctance is to ampere-turns, what (i) resistance is to voltage, (ii) friction is to force, (iii) flux is to mmf, (iv) clouds are to sunshine.
(a) only (i) is true, (b) only (i) and (ii) are true,
(c) only first three are true, (d) only (iii) is false.

Q.214 Reluctance of a magnetic circuit is (i) proportional to the length, (ii) proportional to the inverse of cross-sectional area, (iii) proportional to the magnetic permeability of the path, (iv) inversely proportional to the applied mmf.
(a) only (i) is true, (b) only (i) and (ii) are true,
(c) only first three are true, (d) all are true.

Q.215 Saturation of the magnetic path causes -
(a) decrement of flux,
(b) non-increment of flux,
(c) decrement in rate of rise of flux,
(d) decrement of ampere-turns.

Q.216 In micro sense, flux in a magnetic circuit is produced by -
(a) applied magneto-motive force,
(b) permeance of the magnetic path,
(c) molecules of the material,
(d) magnetic dipoles present in the material.

Q.217 Resistivity of a material is (i) proportional to its length, (ii) inversely proportional to its cross-sectional area, (iii) independent of any dimension.
(a) only (i) and (ii) are true, (b) only (iii) is true,
(c) only first one is true.

Q.218 Retentivity of a magnetic circuit is -
(a) its opposition to flow of flux,
(b) its residual magnetism,
(c) magnetic force required to bring the material to original form,
(d) another name to resistivity.

Q.219 Coercivity of a magnetic circuit is -
(a) its opposition to flow of flux,
(b) its residual magnetism,
(c) magnetic force required to bring the material to original form,
(d) another name to resistivity.

Q.220 For a B-H loop of a magnetic material (i) width at the origin is twice the retentivity, (ii) height at the origin is twice the coercivity.
(a) both of above are true, (b) only one of the above is true,
(c) both are false.

Q.221 B-H loop of a magnetic circuit depends upon (i) magneto-motive force, (ii) quantity of flux flow, (iii) size of the material, (iv) Flux density.
(a) all the above are true, (b) only first three are true,
(c) only first two are true, (d) all are false.

Q.222 In an electrical machine, hysterisis losses are caused by (i) reluctance of the magnetic path, (ii) retentivity of the magnetic material, (iii) coercivity of the magnetic material, (iv) airgaps in the flux path.
(a) only (i) and (ii) are true, (b) only (ii) and (iii) are true,
(c) only (iii) and (iv) are true. (d) only (iv) is false.

Q.223 In an electro-magnetic circuit, eddy current losses are caused by -
(a) flux flow, (b) electromagnetism,
(c) change of flux, (d) change of current.

Q.224 In electrical machines, use of laminated cores (i) reduces hysterisis losses, (ii) reduces eddy current losses, (iii) reduces electical hum, (iv) eases manufacturing.

(a) only first one is true, (b) only (ii) is true,
(c) only (i) and (iii) are false, (d) only last two aretrue.

Q.225 D.C. machines outer body is made of iron-based material because these are -

(a) strong enough, (b) magnetic,
(c) cheap for their strength.

Q.226 A reluctance motor (i) runs at synchronous speed, (ii) is based upon induction principle, (iii) has a cylinderical rotor core, (iv) has very high reluctance.

(a) only (iv) is true, (b) only (iii) and (iv) are true,
(c) only (i) and (ii) are true, (d) only (ii) is true.

Q.227 In a reluctance motor, energy transmission from stator to rotor is by -

(a) electromagnetic induction, (b) magnetic reluctance,
(c) electromagnetism, (d) conductance.

Q.228 In a hysterisis motor, energy flow from stator to rotor is by -

(a) electromagnetic induction, (b) magnetic reluctance,
(c) electromagnetism, (d) conductance.

Q.229 In a hysterisis motor, mechanical power produced is based upon -

(a) material of the rotor. (b) stator core,
(c) rotor current, (d) air gap.

Q.230 A universal motor is basically a -

(a) d.c. shunt motor, (b) d.c. series motor,
(c) induction motor, (d) synchronous motor.

Q.231 Saturation of a magnetic path -

(a) decreases the reluctance, (b) increases the reluctance,
(c) decreases the field strength, (d) increases the flux path.

Q.232 Saturation in magnetic circuits is caused by -

(a) temperature rise of the path,
(b) limited number of the magnetic dipoles,
(c) magnetic hysterisis,
(d) Eddy current flow.

Q.233 Speed of a reluctance motor can be varied by-

(a) input voltage control, (b) relucatnce variation,
(c) frequency control, (d) armature pole changing.

Q.234 A universal motor is so called because it can -
(a) run on any speed, (b) run on any supply,
(c) run on any voltage, (d) be reversed.

Q.235 Universal motors are used in (i) portable machine tools, (ii) domestic mixers, (iii) computers, (iv) ceiling fans.
(a) only (i) is true, (b) only (i) and (ii) are true,
(c) only (iii) is true, (d) only (iv) is true.

Q.236 Fields of (i) d.c. motors (ii) synchronous motors, (iii) d.c. generators, (iv) synchronous generators, can be replaced by permanent magnets.
(a) only (i) and (ii) are true, (b) only (ii) and (iv) are true,
(c) only (i) and (iii) are true, (d) all are true.

Q.237 Stepper motors are similar to (i) variable reluctance motors, (ii) switched reluctance motors, (iii) universal motors, (iv) eddy current motors.
(a) only (i) and (ii) are true, (b) only (ii) and (iv) are true,
(c) only (i) and (iii) are true, (d) all are true.

Q.238 Stepper motor works due to (i) variation of rotor reactance, (ii) variation of rotor reluctance, (iii) variation of armature reactance, (iv) electromagnetic induction.
(a) only (i) and (ii) are true, (b) only (i) and (iii) are true,
(c) only (ii) and (iii) are true, (d) only (iii)and (iv) are true.

Q.239 For high starting torque (i) d.c. series, (ii) slipring induction, (iii) synchronous, (iv) universal, type of motors are used.
(a) only (i) and (ii) are true. (b) only (iii) is false,
(c) only (iv) is false, (d) all are true.

Q.240 For very high speed applications (i) d.c. motors, (ii) induction motors, (iii) universal motors, (iv) synchronous motors are used.
(a) only (i) and (ii) are true, (b) only (i) and (iii) are true,
(c) only (ii) and (iv) are true, (d) only (ii) and (iii) are true.

Q.241 Mostly the following type of generators are used for generation of electricity in large power stations -
(a) induction generators, (b) synchronous generators,
(c) d.c. generators,

Q.242 The most cheap and easy to use type of motor is -
(a) d.c.motor, (b) induction motor,
(c) synchronous motor, (d) universal motor.

Q.243 For arc welding purposes, the following type of generator may be used -

(a) differentially compounded d.c. generator,
(b) synchronous generator,
(c) induction generator,
(d) cumulative compounded d.c. generator.

Q.244 In an induction generator, frequency of generation depends upon (i) number of armature poles, (ii) speed of rotation, (iii) frequency of bus, (iv) winding configuration.
(a) only (i) and (ii) are true, (b) only (iii) is true,
(c) only (iii) is false, (d) all are true.

Q.245 In a synchronous generator, frequency of generation depends upon (i) number of armature poles, (ii) speed of rotation, (iii) frequency of bus, (iv) winding configuration.
(a) only (i) and (ii) are true, (b) only (iii) is true,
(c) only (iii) is false, (d) all are true.

Q.246 A synchronous machine yoke and an induction machine yoke serve the same purpose in the machine provided -
(a) synchronous machine field is on the rotor,
(b) induction motor runs close to synchronous speed,
(c) both machines carry the same load,
(d) both have the same frequency of operation.

Q.247 In electrical terms corona means -
(a) ionization of ambience due to static charge,
(b) sparking at the current carrying conductors,
(c) production of flux around a current-carrying conductor,
(d) discharge of high voltage.

Q.248 In dc machines soft carbon brushes are used for (i) lessen damage to commutator, (ii) low brush-contact drop, (iii) low friction losses.
(a) only (i) is true, (b) only first two are true,
(c) only last two are true, (d) all are true.

Q.249 At what speed must an 8-pole, 50 Hz alternator be run ?
(a) 900 rpm, (b) 750 rpm,
(c) 600 rpm, (d) none of the above.

Q.250 The synchronous machine field ampere-turns are provided considering (i) iron path reluctance, (ii) air gap reluctance, (iii) leakage flux paths. (iv) magnetizing efficiency.
(a) all the above are true, (b) only first three are true,
(c) only first two are true, (d) only first one is true.

Q.251 Large ac motors usually have their windings connected in -
(a) star, (b) delta,
(c) can be star or delta.

ABSTRACT

Law of electromagnetic induction states

When a conductor cuts lines of magnetic flux, an electromotive force is induced in the conductor which is equal to the rate of magnetic flux cutting. If a conductor of length l perpendicular to a magnetic field of flux density B moves transversely with a velocity v, the emf induced in the conductor will be :

$e = Blv$ volt

If the flux is not uniform and sinusoidally varying, represented by :

$B = B_m \cos\theta$

$e = B_m lv \cos\theta$

which represents the rate of change of flux linkage. Hence we can say that the emf induced in a conductor in a magnetic field is equal to the rate of change of flux linking the conductor.

The flux linking a conductor can vary :

(i) due to variation in the magnetic flux, conductor and the field position remaining static. This phenomenon is the basis of transformers, a static electric machine.

(ii) due to the rotation of a steady magnetic flux linking the static coil or rotation of coil in a steady magnetic flux. This phenomenon is the basis of dc and ac rotary machines.

Law of electromagnetic interaction states

When a current-carrying conductor intersects a magnetic field, the conductor experiences a mechanical force. This phenomenon is basis of production of torque in ac and dc motors.

Energy balance

In all types of electromagnetic machines, there are some losses, some of electrical nature and some of mechanical nature. Hence general energy balance is :

Energy input = Energy losses + Energy output

In a transformer, there is no motion hence mechanical losses are zero. Therefore for a transformer :

Electrical energy input = Electrical energy losses + Electrical energy output

In a motor electrical energy is supplied and the mechanical power is produced. The losses are of both types - electrical as well as mechanical, related as follows :

Electrical energy input - Electrical energy losses
= Mechanical energy output + Mechanical energy losses

Similarly for a generator :

Mechanical energy input - Mechanical energy losses = Electrical energy output + Electrical energy losses

6.1. DC MACHINES

Consider a uniform magnetic field of flux density B in a direction and a single-turn rectangular coil with one arm of length l along Z-axis and another arm of length 2r rotating in x-y plane, on its centreline along Z-axis. If the angular velocity of coil is ω radians per second, the flux linked with the coil at any instant is given second, the flux linked with the coil at any instant is given by :

$$A = B\ (2r \cos \omega t)\ l = BA \cos \omega t$$

where A is the area of the coil. Hence voltage induced in the coil :

$$e = \frac{d\phi}{dt} - BA\ \omega \sin \omega t$$

According to Lenz's law, the polarity of this emf will be such that the flux produced by the current due to this voltage will oppose the change of flux. If the coil has N turns :

$$e = -\ NBA\ \omega\ \sin \omega t$$

In dc machine this alternating induced voltage is made unidirectional by a rotary rectifier known as commutator.

6.1.1. DC generators

If the armature winding has a parallel paths and the total number of armature conductors be Z in a p pole machine rotating at n rpm, the generated voltage will be :

$$v_g = \frac{Z}{a}\ p\phi\ \frac{n}{60} = K\phi n$$

where K is a constant for the machine. Thus generated voltage is proportional to speed and field current, if saturation is neglected. A curve showing variation of generated voltage on vertical axis with variation of field current on horizontal axis is called the magnetization curve or the open-circuit characteristic or the saturation curve of the generator.

Generators can be classified as separated excited and self-excited generators. Self-excited generators can further be classified as - shunt field, series field and compound field generators. Shunt field is usually controlled by a series resistance with it while the series field is controlled by diverter resistance in parallel with the field winding or the armature In compound-field generators both the fields can thus be controlled. Compound-field generators are usually of cumulative type.

In all self-excited generators for voltage buildup, some residual magnetism in the correct direction is essential. In addition, for shunt generators shunt-field circuit resistance should be less than the critical resistance. The critical resistance line on the magnetization curve graph coincides with the straight line portion of the magnetization curve. The build voltage is determined by the intersection of the magnetization curve and the shunt field resistance line.

The load current versus terminal voltage characteristic of a generator is called its external characteristics. Shunt generator external characteristic shows a drop in voltage from no load value to full load, given by the relation :

$$V = E - I_L R_a$$

where V = terminal voltage
E = induced emf in the winding
I_L = load current, and
R_a = armature resistance

The characteristic can be modified for terminal voltage rising with the load, droping less with the load or giving terminal voltage value at full-load same as at no head, by providing series field. The stages are respectively called as over-componding, under componding and flat componding and is determined by the ratio of series field ampere turns to shunt-field ampere turns.

6.1.2. DC motors

The torque developed in a dc motor of ϕ field flux per pole and having an armature of conductors carrying a current i_a is given by :

$$T = k\,\phi i_a$$

where k is a constant for the motor.

The voltage drop equation in a motor is :

Applied voltage = Armature drop + Back induced emf

$$V_2 = R_a i_a + kn\phi$$

which gives $n = \dfrac{V_L - R_a i_a}{k\phi}$

DC motors are also classified as shunt-field, series-field and compound-field dc motors.

For shunt motors the speed torque characteristic is of drooping type i.e. as the load torque is increased from no-load to full-load, the speed drops almost linearly. The series motor characteristic is given by :

$$T = kk_f\, i_a^2$$

$$= \frac{k.k_f\, V_L^2}{(R_a + kk_f n)^2}$$

which shows a sharper drop of speed as load torque increases and at zero-load, the motor speed tends to be infinity. Hence a series motor is never run at no-load. The compound motor characteristic is in between series and shunt motor characteristics.

6.2. SYNCHRONOUS MACHINES

A dc-fed field and a rotating magnetic field produced by a usually 3-phase ac armature winding interact to provide torque in a synchronous motor and the induced emf in the armature winding under the influence of a relatively rotating dc field, provides the electrical power generated by a synchronous generator or an alternator.

6.2.1. The alternator

The frequency of the induced emf is given by :

$$f = \frac{pn}{120}$$

where p is the number of poles in the machine and n is its speed in revolutions per minute (rpm).

Alternators are constructed with salient pole rotor or cylindrical rotor. For low-speed water-wheels and low-power diesel generators, salient pole construction is used while for high speed turbo alternators cylindrical-rotor construction is used.

The current-voltage relationship per phase of an alternator armature is given by :

$$V_L = E - I_a (R_a + jX_a)$$

where V_L = terminal voltage vector

E = induced emf vector

I_o = armature current vector

R_a = armature winding resistance, and

X_a = synchronous reactance of the armature winding

6.2.2. The synchronous motor

The 3-phase currents in the armature winding produce a rotating magnetic field of synchronous speed. The static field produced by the excitation of field windings also produce the same number of magnetic poles as those of armature field. The two sets of poles attract each other and are locked to provide the motor a torque at synchronous speed. Hence a synchronous motor has no torque at speeds other than the synchronous speed.

At no-load the armature poles and the field poles axes remain aligned with each other but on load the rotor poles are held back a little by the load from the stator pole and the axes of stator and rotor poles are no more in alignment, but at an angle, known as the power angle. This power angle is positive in motors while in generators this angle is of negative value. Similarly, in generators, induced emf vector leads the terminal voltage vector, while in motors terminal voltage leads the induced emf. Ignoring armature winding resistance, this relation for a synchronous motor is provided by :

$$V_L = jX_s I_a + E$$

The angle between the vectors V_L and E is the power angle δ while the angle between the vectors V_L and I_a is the power factor angle (θ) of the machine. The power input per phase of a synchronous motor is given by :

$$P = V_L I_a \cos \theta$$

$$= \frac{V_L.E}{X_s} .\sin \delta$$

Since E depends upon field excitation, its value can be varied by variation in excitation. This will also vary the power angle δ. If $E > V_L$, we call the motor as over-excited and if $E < V_L$, we call it underexcited. When the motor excitation is increased at a constant load, its power factor is charged from lagging to leading. If the excitation is sufficiently increased, the motor is in a position to supply reactive power to the bus and motor can be used for power factor correction. In this role, a synchronous machine is known as synchronous condenser.

A synchronous motor runs at synchronous speed only, given by :

$$n_s = \frac{120f}{p}$$

From the power equation we find that if the motor load is increased at constant excitation, its power angle also increases with maximum power at $\delta = 90°$. If the load is further increased, the motor torque will not match the load torque and the motor loses synchronism and stops. The torque that causes this situation is known as pull-out torque.

Other types of synchronous motors, where magnetic circuit reluctance variation with angular rotation i.e. saliency, plays important role, are the reluctance motor and the stepper motor.

6.3. POLYPHASE INDUCTION MACHINES

In an induction machine, the stator carries a polyphage, usually a 3-phase, winding which when supplied a 3-phase voltage, produce a rotating magnetic field of speed given by :

$$n_s = \frac{120f}{p}$$

where f is the supply frequency and p is the number of poles of the winding. n_s is called the synchronous speed of the machine.

The rotor of an induction machine carries a short-circuited winding which may be of squirrel-cage type or a usual 3-phase winding. Accordingly the two types of induction machines are known as squirrel-cage induction machine and a wound-rotor induction machine. The number of poles of a wound-rotor are fixed while a squirrel-cage rotor can work with any number of poles. The operating speed of an induction machine is never equal to its synchronous speed. The difference in synchronous speed and the operating speed in per unit value of synchronous speed is known as slip of machine, given by :

$$s = \frac{n_s - n}{n_s}$$

where n is the operating speed.

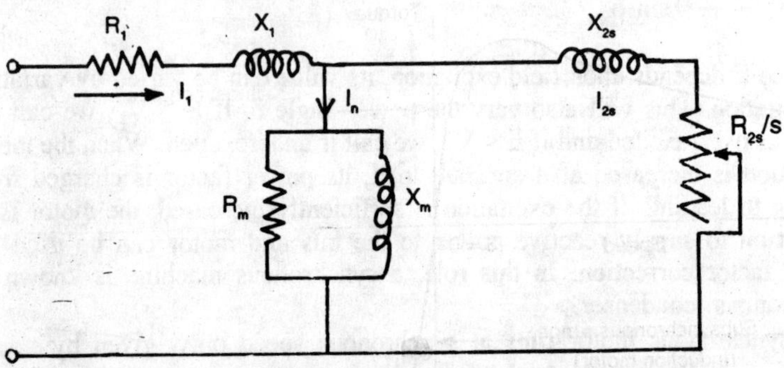

Fig. 6.1. Induction machine equivalent circuit.

Equivalent circuit

The equivalent circuit per phase of an induction machine is shown in Fig. 6.1 where R_1 and X_1 are the stator resistance and leakage reactances and R_{2s} and X_{2s} are the rotor resistance and leakage reactance values referred to stator. The R_m, X_m combination represents the magnetizing circuit of machine, where R_m represents the iron losses in the core of the machine and X_m is for magnetizing current required for magnetizing the magnetic circuit of the machine and maintaining the airgap flux.

Neglecting magnetizing current, $I_1 = I_{2s}$

From the equivalent circuit, we can get the power output of the machine and by dividing it by speed, machine's torque can be evaluated, which comes out to be :

$$T = \frac{3}{\omega_s} \frac{V^2 (R_{2s}/s)}{(R_1 + R_{2s}/s)^2 + (X_1 + X_{2s})^2}$$

from which, we can infer :

(i) The torque speed curve of the machine is shown in Fig. 6.2.

(ii) For normal operations at low values of slip and low values of R_1 and $(X_1 + X_{2s})$ as compared to R_{2s}/s, the torque can be approximated to :

$$T = \frac{3sV^2}{\omega_s R_{2s}}$$

(iii) At starting when s = 1 and ignoring resistance values as compared to reactances, the torque can be approximated as :

$$T = \frac{3V^2 R_{2s}}{\omega_s (X_1 + X_{2s})}$$

Hence the starting torque can be enhanced by increasing rotor resistances of the machine.

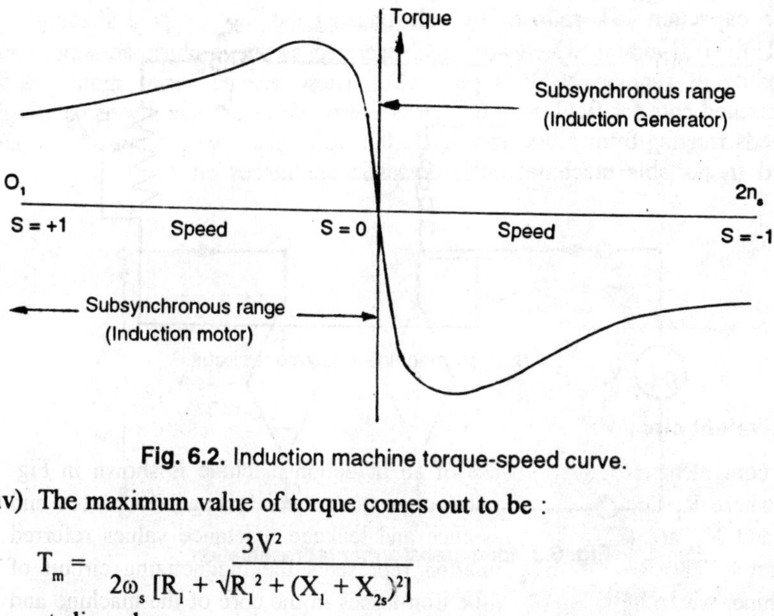

Fig. 6.2. Induction machine torque-speed curve.

(iv) The maximum value of torque comes out to be :

$$T_m = \frac{3V^2}{2\omega_s [R_1 + \sqrt{R_1^2 + (X_1 + X_{2s})^2}]}$$

at slip value :

$$s = \frac{R_{2s}}{\sqrt{[R_1^2 + (X_1 + X_{25})^2]}}$$

which show that the breakdown torque is independent of rotor resistance value, hence by varying rotor assistance only slip of maximum torque is varied.

(v) For negative values of slip the torque also becomes negative, i.e. if the machine is driven at supersynchronous speed, it will work as an induction generator.

6.4. SINGLE-PHASE AC MOTORS

The most commonly used single-phase motors are of induction and universal motors, single phase reluctance motors which are synchronous motors are also sometimes used in low-power fixed-speed applications. Single-phase induction motors have various varieties such as capacitor type, shaded pole type etc.

The main problem in single phase machines arises because a single-phase winding can not produce a rotating magnetic field, which is so far a must for ac motor operations. The magnetic field produced by a single phase winding is stationary in space and pulsating in time, which is equivalent to two rotating magnetic fields of equal magnitudes but rotating in opposite directions. Hence once a single phase motor is started the pulsating magnetic field can be utilized to develop motoring torque. This starting problem is solved by split phasing another single-phase winding on the stator by the help of a capacitor, R/L ratio or by split-phasing the flux by pole shading.

Universal motors which are usual operated as single phase ac supply are capable of running at dc supply also. These are dc series motors with laminated core for field magnetic circuit also. These are usually of very high speeds ranging from 5,000 rpm to 25,000 rpm, hence very compact and are used in portable machine tools, domestic appliances etc.

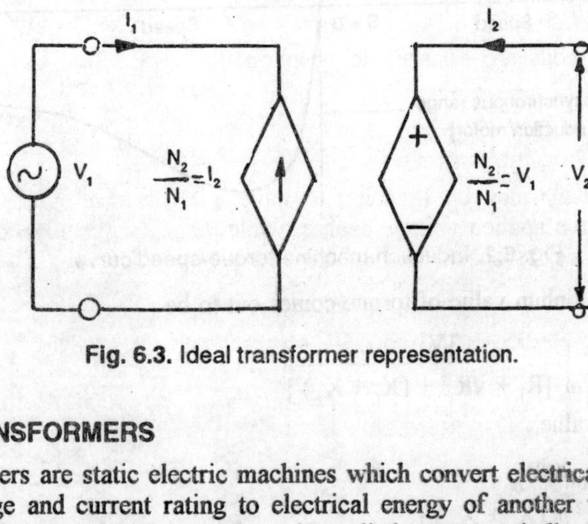

Fig. 6.3. Ideal transformer representation.

6.5. TRANSFORMERS

Transformers are static electric machines which convert electrical energy at one voltage and current rating to electrical energy of another voltage and current rating. An alternating voltage is applied across a winding coil around a magnetic core, thereby producing an alternating flux in the core. This alternating flux, when linked with another coil induces an electromotive force in it in accordance with law of electromagnetic induction. The polarity of the induced voltage is determined by the Lenz's law which states - 'the polarity of the voltage induced by a changing flux tends to oppose the change in flux that induced the voltage'.

The magnitude of the induced voltages in both the coils are proportional to the number of turns in them, i.e. :

$$E_1 = N_1 \frac{d\phi}{dt} \quad \text{and} \quad E_2 = N_2 \frac{d\phi}{dt}$$

If V_1 and V_2 are the terminal voltages at the two coils, I_1 and I_2 be the currents, R_1 and R_2 be the resistances and X_1 and X_2 be the leakage reactances of the two windings, we have the relations :

$$V_1 = E_1 + I_1 (R_1 + jX_1)$$

and $V_2 = E_2 - I_2 (R_2 + jX_2)$

In terms of inductances :

$$E_1 = j\omega L_1 I_1 + j\omega M I_2$$

$$E_2 = j\omega L_2 I_2 + j\omega M I_1$$

ignoring the resistive drops in the windings.

For a usual transformer, if k be the coefficient between the two windings:

$$k = \frac{M}{\sqrt{L_1 L_2}}$$

where $0 \leq k \leq 1$

ideally, the coils are assumed to be perfectly coupled i.e., $k = 1$

$$\frac{E_1}{E_2} = \frac{N_1}{N_2} = \frac{I_2}{I_1}$$

The primary may be assumed to have a dependent current source in addition to the applied voltage source, while the secondary has a dependent voltage source, as shown in Fig. 6.3.

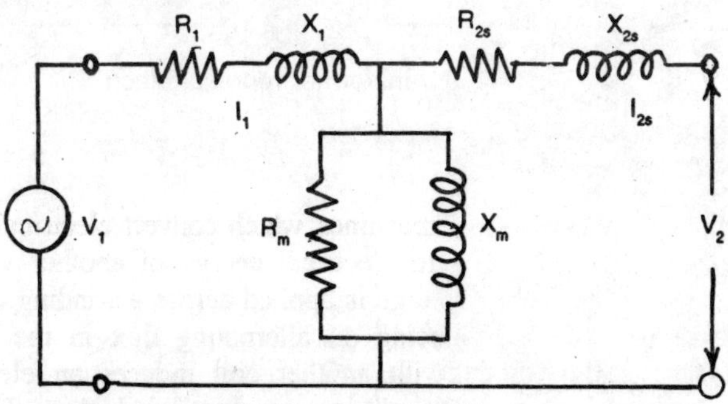

Fig. 6.4. Practical transformer equivalent circuit.

Practical equivalent circuit

The practical equivalent circuit of a transformer is shown in Fig. 6.4, where R_1 and X_1 are the primary resistance and leakage reactance values, R_{2s} and X_{2s} are the secondary resistance as leakage reactance values and R_m and X_m corresponds to the core losses and the magnetizing current.

This circuit can be used to calculate efficiency and voltage regulation of the transformer, which are given by :

$$\text{Efficiency} = \frac{\text{Output power}}{\text{Input power}} = 1 - \frac{\text{Power losses}}{\text{Input power}}$$

$$\text{Voltage regulation} = \frac{\text{(Output voltage at no-load) - (Output voltage at full-load)}}{\text{(output voltage of full-load)}}$$

Electric Drives and Traction

Q.1 An electric drive is supposed to provide motion to the load through the functions of (i) controlling, (ii) modulating, and (iii) converting the electrical power.

(a) all the above are true, (b) only first two are true,
(c) only last two are true, (d) none of the above is true.

Q.2 Electrical drives are said to be four-quadrant drives. This means their -

(a) current and voltage are reversible,
(b) power and torque are reversible,
(c) speed and torque are reversible,.
(d) none of the above.

Q.3 An electrical converter converts -

(a) voltage to current, (b) electrical to mechanical,
(c) dc to ac, (d) none of the above.

Q.4 A cycloconverter in electrical engg. is a device that converts -

(a) one frequency to another, (b) electrical to mechanical,
(c) dc to ac, (d) none of the above.

Q.5 If a dc motor is used in an electrical drive, the power source of the drive -

(a) has to be dc, (b) can be ac or dc,.
(c) can not be ac, (d) none of the above.

Q.6 The maximum speed possible with ac motors in electrical drives is -

(a) 3,000 rpm, (b) 3,600 rpm,
(c) unlimited, (d) none of the above.

Q.7 Static torque or dead torque of a mechanical load is that -

(a) remains due to the static inertia of the load,.

(b) felt only when the drive is not moving,
(c) remains even when the drive is not working,
(d) none of the above.

Q.8 Dynamic torque of a mechanical load is that -
(a) remains same throughout the operation,
(b) felt only when the drive is moving,.
(c) remains even when the drive is not working,
(d) none of the above.

Q.9 A mechanical load driven by an electrical drive will accelerate if -
(a) drive torque > static torque + dynamic torque,
(b) drive torque > static torque,.
(c) drive torque > dynamic torque,
(d) none of the above.

Q.10 If v is the speed of a driven load and a, b are constants, the total friction torque of a load can be written as -
(a) av, (b) av + bv^2,
(c) a + bv, (d) none of the above..

Q.11 Coulomb friction of a load -
(a) constant,. (b) proportional to the speed,
(c) proportional to acceleration, (d) none of the above.

Q.12 Windage friction of a drive is -
(a) constant,
(b) proportional to the speed,
(c) proportional to the square of speed,.
(d) none of the above.

Q.13 Viscous friction of a drive is -
(a) constant,
(b) proportional to the speed,.
(c) proportional to the square of speed,
(d) none of the above.

Q.14 Static friction of a drive is -
(a) greater than that at motion,
(b) greater than that at slight motion,.
(c) lower than that at motion,
(d) none of the above.

Q.15 Load torque which has tendency to drive the motor if the latter's torque is reduced, are called -
(a) gravitational torque, (b) active torque,.
(c) passive torque, (d) none of the above.

Q.16 Load torque which has no tendency to drive the motor even if the latter's torque is reduced, are called -

(a) gravitational torque, (b) active torque,
(c) passive torque,. (d) none of the above.

Q.17 Active torque is characterized by a sign for its quantity. This sign -

(a) changes according to the motor torque direction,
(b) does not change with the motor torque direction,.

Q.18 Passive torque works -

(a) in the direction of the motor torque,
(b) opposite the direction of the motor torque,.
(c) in a fixed direction,
(d) none of the above.

Q.19 A driven load attains a steady speed when -

(a) motor torque capacity = load torque,
(b) motor torque = load torque,.
(c) motor power = load power,
(d) none of the above.

Q.20 A drive is said to be in stable equilibrium when a slight increase in speed makes -

(a) load torque > motor torque,.
(b) load torque < motor torque,
(c) load torquue = motor torque,
(d) none of the above.

Q.21 A drive is said to be in stable equilibrium when a slight decrease in speed makes -

(a) load torque > motor torque,
(b) load torque < motor torque,.
(c) load torquue = motor torque,
(d) none of the above.

Q.22 A flywheel in a drive does the 'load equalization' function, which means keeping -

(a) the drive torque equal to load torque,
(b) electrical load constant under fluctuating mechanical loads,
(c) smmothening the fluctuating mechanical load,.
(d) none of the above.

Q.23 A flywheel can be used with -

(a) constant torque loads, (b) constant speed loads,.
(c) reversible load, (d) all types of loads.

Q.24 A constant torque drive is that which -

(a) provides a fixed torque to the load,
(b) has a fixed maximum torque capability at specified speed,
(c) has a fixed maximum torque capability at whole speed range,.
(d) none of the above.

Q.25 A constant power drive is that which -
(a) provides a fixed power to the load,
(b) has a fixed maximum power capability at specified speed,
(c) has a fixed maximum torque capability at whole speed range,.
(d) none of the above.

Q.26 A constant speed drive is that which -
(a) provides a fixed speed to the load,.
(b) has a fixed maximum speed capability at specified torque,
(c) has a fixed maximum speed capability at whole torque range,
(d) none of the above.

Q.27 A drive's ultimate parameter of fitness is its capability to provide desired -
(a) speed,. (b) torque,
(c) power, (d) none of the above.

Q.28 Speed regulation of a drive is the difference in its speed from no-load to full-load as percentage of -
(a) no-load speed, (b) full-load speed,.
(c) prefixed standard speed, (d) none of the above.

Q.29 The feedback loop components of a drive are a part of the drive -
(a) modulator, (b) controller,.
(c) modulator or controller, (d) none of the above.

Q.30 The feedback loop in an electrical drive can contribute towards the control of (i) torque, (ii) speed, (iii) power, (iv) speed-torque characteristics.
(a) all the above are true,. (b) only first three are true,
(c) only first two are true, (d) none of the above is true.

Q.31 Feedback loop in an electrical drive is provided for (i) protection, (ii) response speed, (iii) accuracy, (iv) mechanical load control.
(a) all the above are true, (b) only first three are true,.
(c) only first two are true, (d) none of the above is true.

Q.32 In a multi-motor drive, the speed synchronization may be obtained (i) either electrically or mechanically, (ii) both electrically and mechanically.
(a) all the above are true,. (b) only (i) is true,
(c) only (ii) s true.

Q.33 If a motor runs with load current I_1 for a duration of t_1 and current I_2 for duration t_2, then for the total duration of $t = (t_1 + t_2)$ the equivalent load current of the motor is -

(a) $(I_1.t_1 + I_2.t_2)/t$, (b) $(I_1 + I_2)/2$,

(c) $[(I_1)^2 + (I_2)^2]/2$, (d) $[(I_1)^2.t_1 + (I_2)^2.t_2]/t$,.

Q.34 For optimum utilization of a motor, the limitation to be considered is -

(a) its current rating, (b) power rating,

(c) torque rating, (d) thermal rating..

Q.35 Continuously-variable speed control drives are mostly based upon the use of -

(a) frequency-controlled ac motors,

(b) dc motors,.

(c) universal motors,

(d) none of the above.

Q.36 In a dc motor, other parameters remaining fixed, torque is proportional to (i) armature voltage, (ii) armature current, (iii) field current.

(a) all the above are true, (b) only (i) and (ii) are true,

(c) only (ii) is false, (d) only (i) is false..

Q.37 In a dc motor, other parameters remaining fixed, speed is proportional to (i) armature voltage, (ii) armature current, (iii) inverse of field current.

(a) all the above are true, (b) only (i) and (ii) are true,

(c) only (ii) is false, (d) only (i) is false.

Q.38 In constant-torque mode swpeed of a dc motor can be controlled by -

(a) armature voltage control, (b) field current control,

(c) armature current control, (d) none of the above.

Q.39 In constant-power mode swpeed of a dc motor can be controlled by -.

(a) armature voltage control, (b) field current control,.

(c) armature current control, (d) none of the above.

Q.40 In a dc motor drive using thyristors and diodes, power flow in both the directions is possible with (i) full-wave converter, (ii) fully-controlled converter, (iii) half-controlled converter.

(a) all the above are true, (b) only (i) is false,

(c) only (iii) is false, (d) only (ii) is true..

Q.41 A dc motor can be fed from a dc source of constant voltage, with electronic armature voltage control through -

(a) cycloconverter, (b) inverter and rectifier,
(c) chopper circuit, (d) none of the above.

Q.42 With the use of a plain chopper circuit, dc motors can have -
(a) four-quadrant operation, (b) two-quadrant operation,.

Q.43 With a 3-phase unbalanced supply to an induction motor, the net torque is less than the usual torque because of -
(a) circulating currents,
(b) negative phase sequence voltages,.
(c) harmonics,
(d) none of the above.

Q.44 Unbalance in the rotor circuit of an induction motor may cause a dip of torque at half the synchronous speed, due to the presence of -
(a) higher rotor resistance,
(b) unbanced loadingof stator,
(c) negative sequence currents in te rotor,.
(d) none of the above.

Q.45 When an induction motor is supplied from a nonsinusoidal power source, the motor efficiency goes down due to -
(a) higher losses due to harmonic components,
(b) skin effect at higher harmonics,
(c) both (a) and (b),.
(d) none of the above.

Q.46 Starting method of a motor is chosen based upon the (i) load starting torque requirements, (ii) power supply stiffness, (iii) response speed.
(a) all the above are true, (b) only first two are true,
(c) only last two are true, (d) none of the above.

Q.47 As compared to a direct-on-line starter, a star-delta starter for starting an induction motor, reduces the starting current to -
(a) 1/2, (b) $1\sqrt{3}$,.
(c) $1/\sqrt{2}$, (d) none of the above.

Q.48 In autotransformer starting of an induction motor, if starting voltage is reduced by ratio a, the starting torque is reduced by -
(a) a, (b) $\sqrt{a}$,
(c) a^2, (d) none of the above.

Q.49 In a saturable-core reactor used for induction motor starting, the saturation of the core causes -
(a) increment of impedance, (b) decrement of impedance,.

(c) chopping the current waves, (d) none of the above.

Q.50 Regenerative braking can be applied in induction motors, by -
(a) reversing the supply terminals,
(b) reducing the supply voltage,
(c) lowering the input frequency,.
(d) all the above.

Q.51 Plugging as a method of braking can be applied in induction motors, by -
(a) reversing the supply terminals,.
(b) reducing the supply voltage,
(c) lowering the input frequency,
(d) all the above.

Q.51 While braking a motor by plugging, when the motor comes to zero speed, its power supply should be cutoff else -
(a) the motor will again start in forward direction,
(b) the motor will run in reverse direction,.
(c) motor will get ceased,
(d) none of the above.

Q.52 When the rotation of a moto is utilized to generate power and dissipate it for braking, this type of braking is called -
(a) plugging, (b) regenrative braking,
(c) dynamic braking, (d) none of the above.

Q.53 In induction motors ac dynamic braking is achieved by -
(a) reversing any two phase connections,
(b) disconnecting from supply and shorting the terminals,
(c) disconnecting one phase from the supply,.
(d) none of the above.

Q.54 For dc dynamic braking of induction motors, dc is injected into the motor winding -
(a) with normal ac supply, (b) with reversed ac supply,
(c) with ac supply disconnected, (d) none of the above.

Q.55 In zero-sequence dynamic braking of an induction motor, the three phases of the stator winding are supply with ac/dc currents with -
(a) equal magnitudes and phases,.
(b) unequal magnitude and same phases,
(c) equal magnitude and different phases,
(d) none of the above.

Q.56 In a six pole induction motor stator, each phase winding has -
(a) two poles, (b) three poles,

(c) six poles, (d) none of the above.

Q.57 In speed control of induction motors through pole changing, the obtainable speed ratios are -

(a) only integers, (b) any positive value,
(c) any value, (d) none of the above.

Q.58 The pole changing method of induction motor speed control is usually applicable to sq. cage motors because -

(a) slipring stator poles can't be changed,
(b) slipring rotor poles get changed with stator pole change,
(c) sq. cage rotor can have any number of poles,.
(d) none of the above.

Q.59 In a fan drive, the load power is proportional to -

(a) speed, (b) speed-square,
(c) speed-cube, (d) none of the above.

Q.60 Stator voltage control method of induction motor speed control can be applied only to loads whose -

(a) torque varies with speed, (b) torque remains constant,
(c) power varies with speed, (d) none of the above.

Q.61 Voltage control of induction motors for speed control can be achieved electronically by the use of -

(a) converters,
(b) cycloconverters,
(c) thyristors in anti-parallel or triacs,.
(d) none of the above.

Q.62 Induction motor speed control can be achieved in constant-torque mode by changing input voltage (V) and frequency (f) keeping -

(a) V/f constant,. (b) V constant,
(c) V^2/f constant, (d) none of the above.

Q.63 Induction motor speed control can be achieved in constant-power mode by changing input voltage (V) and frequency (f) keeping -

(a) V/f constant, (b) V constant,.
(c) V^2/f constant, (d) none of the above.

Q.64 In V/f constant operation for induction motor speed control, V/f ratio is increased at low speeds for -

(a) compensating stator resistance drop,.
(b) compensating harmonics,
(c) compensating negative sequence currents,
(d) none of the above.

Q.65 Optimum operation of induction motor with frequency-control is achieved by keeping V/f constant, because, with this condition-

(a) motor power remains constant,
(b) motor flux remainsconstant,.
(c) motor speed remains constant,
(d) none of the above.

Q.66 In eddy-current couplings, the coupling effect is achieved by -

(a) dc current supply, (b) ac current supply,
(c) ac and dc both, (d) none of the above.

Q.67 If a fixed resistor R and an on/off transistor circuit with duty ratio d, are connected in parallel, the effective resistance will be-

(a) d × R, (b) (1 – d)/R,
(c) (1 – d)R, (d) none of the above.

Q.68 In rotor resistance control for slipring motor speed control, for the best performance,, the three phases of rotor -

(a) should be equally loaded, (b) may be loaded unequally,
(c) should be loaded unequally, (d) none of the above.

Q.69 For speed control of induction motors, through the use of power injection in the rotor, the rotor power injection -

(a) is directly done from mains,
(b) is required to be at slip frequency,
(c) isrequired to be at fixed voltage,
(d) none of the above.

Q.70 In static Scherbius drive for speed control of induction motors, the regulation is done of rotor -

(a) frequency, (b) resistance,
(c) rectified dc power, (d) none of the above.

Q.71 In static Kramer drive for speed control of induction motors, the regulation is done of rotor -

(a) frequency, (b) resistance,
(c) rectified dc power, (d) none of the above.

Q.72 Synchronous motors can not be braked by -

(a) dynamic braing, (b) plugging,
(c) regenerative braking, (d) friction braking,

Q.73 Synchronous motor speed can be controlled by (i) input voltage control, (ii) field current control, (iii) input freuency control.

(a) all the above are true, (b) only first two are wrong,
(c) only last two are true, (d) none of the above is true.

Q.74 Brushless dc motor is -

(a) dc motor with permanent magnets,
(b) inverter-fed synchronous motor with permanent magnets,.
(c) induction motor with permanent magnets,
(d) none of the above.

Q.75 In a brushless dc motor, the inbuilt inverter transistors are controlled by -
(a) synchronous pulse generator, (b) rotor position sensors,.
(c) programmed pulse generator, (d) none of the above.

Q.76 Speed of a brushless dc motor can be controlled by -
(a) input voltage control, (b) inverter frequency control,
(c) load control, (d) none of the above.

Q.77 A stepper motor rotor rotates by a fixed predecided angle on the application of -
(a) each increment of input voltage,
(b) each reversal of input voltage,
(c) each pulse of input voltage,.
(d) none of the above.

Q.78 The speed of a stepper motor can becontrolled by -
(a) input voltage control, (b) rate of pulse supply,.
(c) load control, (d) none of the above.

Q.79 The basic principle of operation of stepper motor is -
(a) alignment in the direction of minimum reluctance,.
(b) magnetic attraction,
(c) magnetic repulsion,
(d) none of the above.

Q.80 The permanent magnet stepper motors, as compared to the variable reluctance stepper motors, have -
(a) higher torque /stator current ratio,
(b) higher stepping rate,
(c) lower rotor inertia,
(d) none of the above.

Q.81 The variable-reluctance stepper motors, as compared to permanent-magnet stepper motors have -
(a) higher torque /stator current ratio,
(b) higher stepping rate,
(c) higher rotor inertia,
(d) none of the above.

Q.82 In a hybrid stepper motor, torque is developed due to -
(a) permanent magnets on therotor,
(b) variable reluctance of the rotor,

(c) both (a) and (b), (d) none (a) or (b).

Q.83 Presently manufactured solar cells have efficiency in the range -

(a) 10 - 16 percent,. (b) 20 - 30 percent,
(c) 30 - 50 percent, (d) none of the above.

Q.84 For manufacturing solar cells, the material used is -

(a) selenium, (b) silicon,.
(c) germanium, (d) none of the above.

Q.85 Presently manufactured, each solar cell provides the following voltage anf current range -

(a) 0.5 - 0.8 V, 1 - 2 A, (b) 0.5 - 0.8 V, 2 - 3 A,.
(c) 0.2 - 0.4 V, 2 - 3 A, (d) none of the above.

Q.86 Solar powered pumps below 1 kw rating are preferred to be driven by -

(a) dc motors, (b) induction motors,
(c) synchronous motors, (d) none of the above.

Q.87 Solar powered pumps above 1 kw rating are preferred to be driven by -

(a) dc motors, (b) induction motors,.
(c) synchronous motors, (d) none of the above.

Q.88 With solar panels, use of a battery is -

(a) essential, (b) desirable,.
(c) not desirable, (d) none of the above.

Q.89 An electrical multiple unit (emu) train is that -

(a) whose each coach is separately motorized,
(b) is driven by multiple motors attached to various coaches,.
(c) has one drive and multiple coaches,
(d) none of the above.

Q.90 The normal voltage selected for electric traction is -

(a) 2,500 ac, (b) 25,000 dc,
(c) 25,000 ac, (d) none of the above.

Q.91 For underground trains, voltage is between -

(a) 2,000 - 3,000 volts ac, (b) 500 - 1,000 volts ac,
(c) 500 - 1,000 volts dc, (d) none of the above.

Q.92 Coefficient of adhesion for a traction system is given by the following ratio, for traction without slipping -

(a) maximum driven load/load of the driving coach,
(b) maximum tractive effort/ weight of the train,
(c) maximum tractive effort/ weight on driving axle,.
(d) none of the above.

Q.93 In a traction system, the coefficient of adhesion -

(a) increases as speed increases, (b) increases as speed decreases,.
(c) is independent of speed, (d) none of the above.

Q.94 In a traction system, coefficient of adhesion is -

(a) higher for series-connected motors than that for parallel,
(b) lower for series connected motors than that for parallel,.
(c) independent of motor connections,
(d) none of the above.

Q.95 In a traction system, as compared to stepped control, a stepless control -

(a) provides better coefficient of adhesion,.
(b) provides lower coefficient of adhesin,
(c) does not affect coefficient of adhesion,
(d) none of the above.

Q.96 In an emu traction, motor connected to a newer (larger) driving wheel, carries more load than that connected to older wheels.

(a) true, (b) false,

Q.97 Due to wear ter of driving wheels in an emu traction system, the load sharing between various motors is more uniform in -

(a) dc series motors, (b) dc separately-excited motors,
(c) both (a) and (b), (d) none of the above.

Q.98 If a, b, c are constants, tractive effort of a traction system running with speed V, can in general, be denoted by -

(a) $a + bV$, (b) $a + bV^2$,
(c) $a + bv + cV^2$, (d) none of the above.

Q.99 Specific energy consumption of a traction system is energy consumed in watt-hours -

(a) per ton of train weight per kilometer distance travelled,.
(b) per kilometer of distance travelled,
(c) per ton of train load,
(d) none of the above.

Q.100 In traction voltage source inverters (VSI) are used to feed -

(a) dc series motors,
(b) dc separately excited motors,
(c) squirrel cage induction motors,
(d) synchronous motors.

Q.101 In traction load commutated inverters (LCI) are used to feed -

(a) dc series motors, (b) dc separately excited motors,
(c) squirrel cage induction motors,
(d) synchronous motors..

ABSTRACT

Electric drive means a system to provide mechanical motion to a physical mass by utilizing power from the electric supply system. Hence an electric drive includes :

(i) Electric power switching devices,
(ii) Power transmission devices,
(iii) Power control equipment and circuitary,
(iv) Electric motor.

The most important of these are the power control devices and the electric motors. Electric motors have been studied in another chapter. Hence this chapter will concentrate on the electric power control devices to the motor to run a load.

Mechanical loads may be rotational such as pumps, compressors, etc. or translational such as trains rolling stock. So far rotary motors have no competition from linear motors, and are ruling the field. Hence most of the loads are supplied with rotational torque to drive them.

The rotational torque supplied to a mechanical load is required to overcome :

(i) dead load,
(ii) inertia,
(iii) acceleration,
(iv) friction,
(v) energy losses.

7.1. MECHANICS OF LOAD

(a) Translational motion

Force necessary for acceleration $F = m\alpha = m.\frac{dv}{dt}$

where m is the mass of the body, α is the acceleration and v is the linear velocity.

(b) Rotational motion

Torque necessary for acceleration $M = J.\alpha = J\frac{d\omega}{dt}$

where J is the moment of inertia, α is the angular acceleration and ω is the angular velocity.

Apart from the accelerating torque, if there is some torque in the opposite direction also, such as friction, the equation becomes :

$$M_a - M_r = J\alpha = J\frac{d\omega}{dt} = J\frac{d^2\theta}{dt^2}$$

where θ is the angular displacement

M_a - accelerating torque, and

M_r - resistant torque.

Torque is also known as moment of force.

If the rotating body has a radius r where its mass m can be treated as concentrated, its moment of inertia :

$J = mr^2$

The work done in angular displacement θ

$W = M.\theta$

and power $P = \frac{d\omega}{dt} = M \frac{d\theta}{dt} = M\omega$

Energy consumed from angular velocity ω_1 to ω_2

$\Delta W = ½J(\omega_2^2 - \omega_1^2)$

Kinetic energy $= ½J\omega^2$

7.2. DC MOTOR DRIVES

7.2.1. Starting methods

(a) Full voltage starting

In shunt and compound motor starting, care is taken to first feed the field winding and than only energize the armature. Small motors may be started with full armature voltage at the starting itself. This creates serious problems of dangerous sparks at the commutator, overheating of armature winding, high starting current and high acceleration torque. Hence this method is avoided in most of the situations of starting large motors.

(b) Armature resistance starting

This is an old classical method. A specially designed starter which varies the armature series resistance in steps is used. The starter has the provision of not feeding armature voltage unless field is energized.

Another variation of this method is to gradually reduce the armature series resistance in steps through the magnetic contactors. This provides automatic starting of the motor.

7.2.2. Speed control

Speed of a DC motor can be controlled either by field variation (above base speed) or by armature voltage variation (below base speed) or by variation of both.

(i) Armature voltage control

This may be achieved by any of the following three methods :

(a) Armature series resistance

By introducing a resistor with position for variation, in series with the rotor armature the net voltage supplied to the armature can be controlled. This is a simple method but involves loss of power in the series resistors. By this method shunt motor characteristics of torque versus speed have a steeper droop for higher resistance values.

(b) Ward Leonard control

In this method, the motor armature is fed from the output of a dc generator where voltage is controlled by the control of field excitation of the generator. Thus the loss of power in the resistors is avoided and a smooth control of speed is achieved. This requires bulky motor generator set to supply the motor armature. In present times, this method is also avoided on cost and space grounds.

(c) Solid-state control

By the use of thyristors smoothly variable dc supply can be obtained from the 3-phase or 1-phase ac supply. This method is a modern favoured method of speed control. By feedback control techniques, stabilized speed, limited current and smooth variation of speed are achievable by this method.

But, since thyristor-controlled dc supply has harmonic components in it, derating of the motor has to be done with this method of speed control.

(ii) Field control

DC motor speed varies in inverse proportion of field current if armature voltage is kept constant. Hence by excitation control, motor speed can be controlled. In this method as the excitation is reduced, motor torque is also reduced proportionally with the speed. Hence the motor operates with constant horsepower mode, since the excitation current can not be increased beyond rated value, this method of speed control is applicable to provide speeds above base speed of the motor.

7.3. INDUCTION MOTOR DRIVES

Power balance

(i) Motor operation

Stator input power - Stator (copper + iron) losses = Electromagnetic power
= Rotor output power + Rotor (Copper + Iron + Mechanical) losses

(ii) Generator operation

Rotor input power - Rotor (Iron + mechanical + copper) losses
= Electromagnetic power
= Stator output power + stator (copper + iron) losses

(iii) Braking operation

[Rotor input power - Rotor (mechanical + iron) losses]

+ [Stator input power - Stator (Copper + Iron) losses]
= Rotor copper losses

7.3.1. Motor starting

(i) Direct-on-line (DOL) starting

For usual motors :

Motor capacity (kw)	*1 to 5.5*	*67 to 30*	*30 to 40*	*40 to 100*
Starting current % of rated current	720	720	650	650
Starting torque % of rated torque	160	120	110	90

Motor is supplied with full ac supply voltage through magnetic contactor.

(ii) Star/Delta starting

At start motor stator 3-phase windings are connected in and full line voltage is supplied. When motor reaches nearly full speed, stator phase winding connections are changed to delta for full motor power output.

This reduces starting current to almost one third of the direct-on-line starting current.

(iii) Stator reactor starting

A reactor is connected in series with the stator winding at the time of starting the motor. When motor picks up speed the reactor is bypassed. In this method if the starting current is reduced by a factor k as compared to the direct-on line starting, the starting torque is reduced by a factor of k^2 as compared to DOL starting. Hence this method is applicable to no-load starting conditions.

(iv) Auto-transformer starting

At start, the stator of the motor is supplied a low voltage through an auto transformer. When the motor picks up speed, the full voltage is supplied by bypassing the autotransformer. In this method, if the starting voltage is reduced by a factor of k, the starting current and the starting torque are reduced by a factor of k^2. Hence this method is also applicable to no-load or light-load starting of the motor.

(v) Rotor resistance starting

For starting wound-rotor motors, a three-phase balanced and variable resistor is connected across the sliprings. The motor stator is fed with the usual line voltage. The starting torque of the motor is high with low starting current. As the motor picks up speed, rotor resistor is gradually reduced and finally short-circuited.

7.3.2. Speed control

(i) Stator-control methods

(a) Variation of voltage supplied to motor

This method reduces the torque of the motor which is proportional to the square of the stator voltage, hence applicable to small motors.

(b) Stator pole-changing

For this stator winding has the provision of stator pole changing by change of terminal connections. By this the synchronous speed and hence the running speed of the motor is controlled in integer multiplies. This method is applicable to squirrel-cage motors only.

(c) Stator supply frequency variation

This is an effective method of smooth speed control. Stator voltage/ frequency is kept constant while varying the frequency, hence motor torque also varies proportional to the speed. This method requires the provision of a costly inverter.

(ii) Slip-power control methods

In wound-rotor induction motors, slip power can be made available at the sliprigs. Slip ring motors have the disadvantages of higher cost and more maintenance. For speed control of these motors, use of resistors in the rotor circuit was done to dissipate slip power. This caused a loss of energy. Recent trend is to control the wound-rotor induction motors through control of slip-power by solid-state devices. The advantages of such methods are as follows:

(i) The additional cost of wound-rotor motors and their maintenance is compensated for by the usefulness of the slip power.

(ii) The speed of the motor can be smoothly controlled.

(iii) The speed of the motor does not remain limited to its usual speed of nearly synchronous speed. By bidirectional control of slip power i.e. from or to the rotor, the motor speed can be controlled in sub synchronous as well as supersynchronous range.

(iv) The slip power control method can be used in variable-speed, constant frequency mode of motor control, such as in wind energy generators and slipboard power supplies.

(v) The control is like that of dc machines, being simple and with minimum of stability problems.

For slip-power control following methods are adopted.

(a) Static control of rotor resistance

Though adding a rotor resistance causes loss of energy, it provides high starting-torque, low starting current and high power factor, hence preferred in some cases. The rotor resistance can be controlled by a static switch in parallel with a high resistance. This method is advantageous in the sense that no bulky resistances are required, variation of a single resistance keeps the three-phases of rotor in balanced condition and smooth variation of resistance and hence the motor speed and torque are possible.

In this method the three-phase step-power from rotor sliprings is rectified to dc by a three-phase full-wave rectifier bridge, the dc voltage then converted to a dc current by a series inductor of large inductance and then the current is passed through a resistor of moderate value.

The resistor is bypassed by a parallel GTO chopper from time-to-time when the chopper is off the current passes through the resistor and when the chopper is on, the resistor is bypassed and a heavy current passes through the thyristor of the chopper, limited only by the series inductor. If the duty cycle of the chopper be d, such that d = on time/time period, the effective resistance of the circuit will be $R_o = R(1 - d)$. The value of d can be varied to vary the effective resistance. This type of circuit is shown in Fig. 7.3.

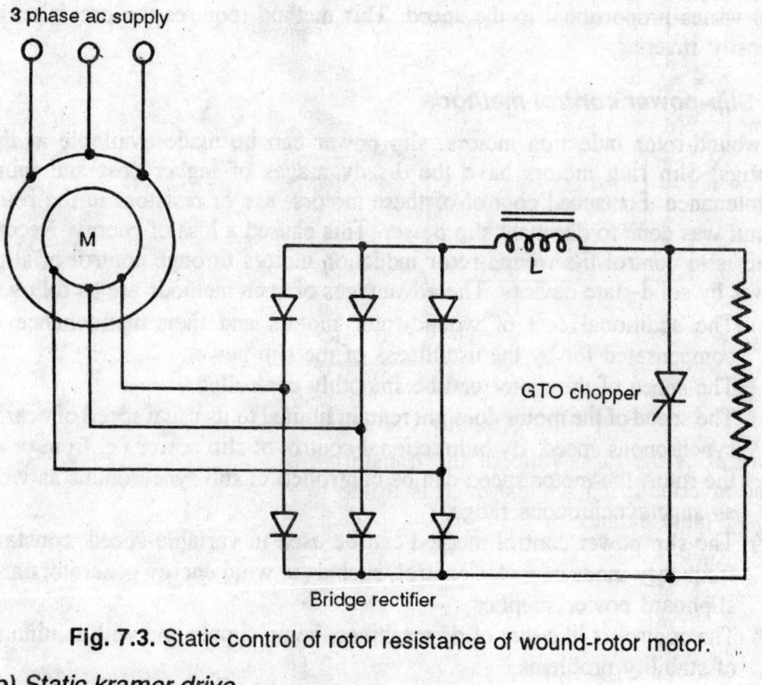

Fig. 7.3. Static control of rotor resistance of wound-rotor motor.

(b) Static kramer drive

In a conventional kramer control slip power is converted into mechanical power and added to the motor output. The same effect of smooth speed control can be achieved by a more reliable static control of slip-power and feeding it back to the supply line. The circuit arrangement is shown in Fig. 7.4.

The slip power from the shipring is rectified to dc and converted to ac by a thyristor inverter. By the gate control of the inverter thyristors the speed of the motor can be smoothly and effectively controlled.

For simplification assume that the stator and rotor drops are negligible. The rectified dc voltage is given by :

$$V_d = \frac{1.35}{n_1} s.V_L$$

where s is the slip of the motor, V_L is the stator line voltage and n_1 is the stator/rotor turns ratio of the motor.

From inverter side, inverter input voltage is given by :

$$V_1 = \frac{1.35}{n_1} V_L (\cos \alpha)$$

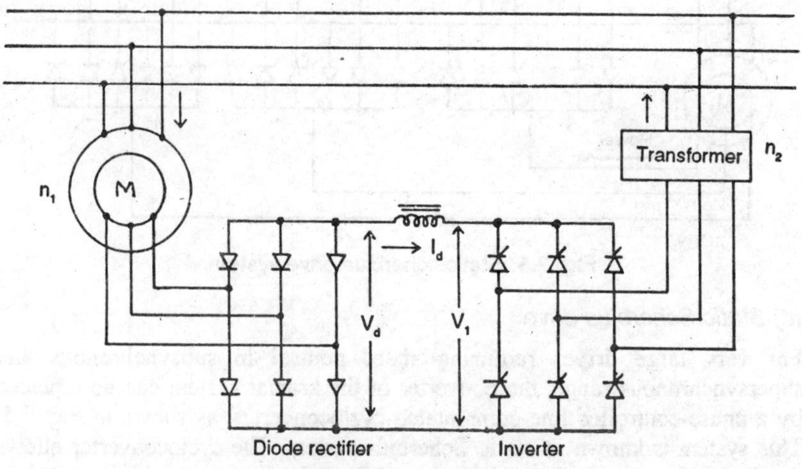

Fig. 7.4. Static Kramar drive system.

where n_2 is the turns ratio of the transformer line side to inverter side, α is the inverter firing angle, which is in the range of 90° to 180°. Since V_d and V_1 must balance :

$$s = \frac{n_1}{n_2} (\cos \alpha)$$

If $n_1 = n_2$

$$w_r = w_s (1 - |\cos a|)$$

where w_r is the motor speed and w_s its synchronous speed. Since cos α varies between 0 and 1, the speed of the motor can be controlled between zero and synchronous speed, by change of firing angle.

If P_g is the airgap power and P_m the mechanical power output of the motor, the slip power :

$sP_g = V_1 I_d$

where I_d is the current in the dc link between the rectifier and the inverter.

Also $P_m = T.w_r = T.w_s\ (1 - s) = (1 - s)\ P_g$

$$\text{Therefore } T = \frac{V_1 I_d}{s.w_s}$$

$$= \frac{1.35\ .\ V_L}{n_1.w_s} I_d$$

which indicates that torque is proportional to I_d. And the slip is proportional to V_d. This is like the control of a dc separately excited dc motor.

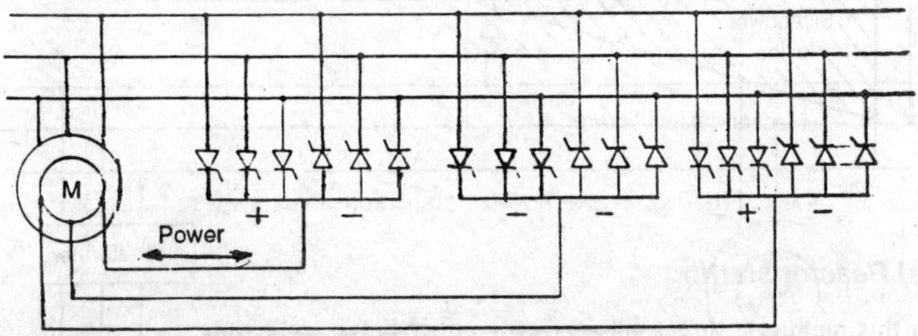

Fig. 7.5. Static scherbius drive system.

(c) Static Scherbius drive

For very large drives requiring speed control in subsynchronous and supersynchronous range, the converter of the kramar system can be replaced by a phase-controlled line-commutated cycloconverter, as shown in Fig. 7.5. This system is known as static Scherbius system. The cycloconverter allows the flow of power in both directions, hence the motor can be operated in four-quadrants of motoring and regenerating in sub-synchronous and super synchronous range, as shown in the power balance diagram of Fig. 7.6.

7.4. SYNCHRONOUS MOTOR DRIVES

7.4.1. Starting methods

The motor is provided with additional windings on the rotor field pole faces to generate induction torque during starting. At synchronous speed this winding becomes ineffective. The winding also suppresses hunting of the machine due to load variations. Such motor are known as induction-synchronous motors.

Since this method is applied to large synchronous motors their stators are not supplied with full line voltage at the starting, else the starting current will be very high. To reduce the stating current either of the two methods are applied :

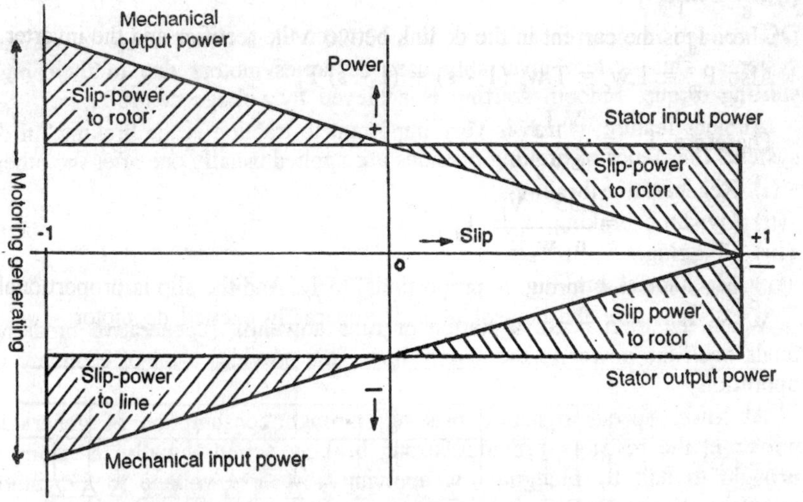

Fig. 7.6. Power flow for static scherbius system.

(a) Reactor starting

In this method a three phase reactor either between the supply and the stator winding or between the stator winding or its short-circuited neutral point, is connected. When the motor picks up near synchronous speed, the reactor is short circuited.

(b) Auto transformer starting

The motor stator is supplied a low voltage at the time of starting through an autotransformer, which is bypassed later.

7.4. TRACTION

Railway traction has a long history starting from steam engines to diesel electric locomotives and now to electric traction. Steam engines had a short span of life and created pollution in the atmosphere due to coal-fired boilers for steam generation. Diesel locomotives are driven by electric motors which are fed from the generators driven by diesel engines on the train. Hence this system is known as Diesel-electric traction. This method proves costlier than the modern method of laying a single wire overhead line alongwith the railway track and feeding the motors driving the vehicle with a supply through this wire, which is fed from ac transformers at various points. The power to the train from the overhead wire is supplied through the use of a pentograph collector at roof of the control cabin. This traction system may be classified on the basis of type of power supply.

(i) DC system

DC voltage ranging from 1500 V to 3000 V has been in use in various systems. This system invariably uses dc series motors due to their high starting torque. Smooth starting is achieved by voltage control.

Another feature, which is very important to traction is the braking. In dc systems four types of braking methods are applied usually one after the other:

(i) Regenerative braking,
(ii) Dynamic braking,
(iii) Plugging,
(iv) Mechanical braking.

When the train nears a station or runs downhill, regenerative breaking feeds back the power to the supply line. This provides a lot of economy in operation.

At lower speeds dynamic braking, through consumption of generated power in the resistors provides strong braking action. Finally the train is brought to halt by plugging i.e. applying a reverse voltage to the motor armature circuit.

Mechanical brakes are provided for safety reasons and to be applied for long period stoppage of the train at a point.

AC system

A single phase A.C. voltage in the range of 11 kv to 16 kv has been used in different systems. Currently AC systems of electric traction are more popular, mainly because of the convenient power supply system, as dc transformation is costlier and inconvenient.

In AC system of traction AC operated universal motors, which are like dc series motors except the field core laminations, are used. These provide same type of control characteristics as dc series motors.

8

Measurements and Instrumentation

Q.1 Basically a transducer is a device that -

(a) senses a parameter,
(b) converts a parameter,
(c) senses and converts a parameter,
(d) none of the above.

Q.2 In a measuremnt system, a transducer must be sense the changes in the object, to which it is connected -

(a) in whole, (b) selectively,
(c) discretely, (d) none of the above.

Q.3 The interaction of a transducer with the object of measurement should be -

(a) one-way, (b) two-way,
(c) one or two way, (d) none of the above.

Q.4 The output of a transducer should be -

(a) proportional to input, (b) a function of input,.
(c) may be unrelated to input, (d) none of the above.

Q.5 While you read a book, the spots on the paper are measured. In this measurement the role of the transducre is performed by -

(a) the book, (b) your eye,
(c) your brain, (d) none of the above.

Q.6 In a measurement system, the signals from a transducer are again converted by the signal conditioning device to make them -

(a) more linear, (b) more sensible,.
(c) larger, (d) none of the above.

Q.7 In measurement systems, some of the signal conditioning devices are (i) amplifiers, (ii) filters, (iii) a.d converters, (iv) integraters.

(a) all the above are true, (b) only first three are true,
(c) only first two are true, (d) none of the above.

Q.8 Null type of measuring instruments basically do the function of a -
(a) adder, (b) adder/substractor,
(c) comparator, (d) none of the above.

Q.9 In measurement systems, instrument-bias is an inherent characteristic of the instrument, provided to it -
(a) by design, (b) by accidental error,
(c) by extra-effort, (d) none of the above.

Q.10 A spring balance provides different weights at differnt altitudes, this type of error in measurement is known as -
(a) instrument error, (b) environmental error,.
(c) external error, (d) none of the above.

Q.11 A loading error in a measurement is caused by the measuring-instrment itself by -
(a) altering the original condititions of the system,.
(b) its inherent bias,
(c) adding human factor in the system,
(d) none of the above.

Q.12 In measurement technology, accuracy of a measuring instrument is defined as -
(a) percentage deviation from true value,.
(b) percentage actual reading aaginst true value,
(c) inverse of deviation in percent of true value,
(d) none of the above.

Q.13 In measurement technology, precision of a mesuring instrument is defined as -
(a) percent deviation in repetitions,.
(b) percent deviation from true value,
(c) percent of actual reading against true value,
(d) none of the above.

Q.14 In measurerment technology, accuracy and precision respectively, are, in fact, indiacative of -
(a) repeatability and deviation,
(b) deeviation and repeatability,
(c) deviation in single and repeated readings,.
(d) none of the above.

Q.15 A precise measurement -
(a) will always be accurate, (b) may not be accurate,.

(c) none of the above.

Q.16 Resolution of a measuring instrument is -

(a) the smallest value it can read,.
(b) the smallest division on its dial,
(c) the maximum value it can safely measure,
(d) none of the above.

Q.17 Threshold of a measuring instument is -

(a) maximum value of input for which no output will be detected,.
(b) smallest value it can read,
(c) smallest division on its dial,
(d) none of the above.

Q.18 Static sensitivity of a measuring instrument is -

(a) ratio of input signal to output signal,.
(b) ratio of output signal to input signal,
(c) minimum input signal that can be measured,
(d) none of the above.

Q.19 The deflection factor or inverse-sensitivity of a measuring instrument is -

(a) ratio of input signal to output signal,
(b) ratio of output signal to input signal,.
(c) minimum input signal that can be measured,
(d) none of the above.

Q.20 Hysterisis of a measuring instrument is the deviation in its readings, when the readings are taken -

(a) in ascending and descending orders,.
(b) its inability to show a minimum input in incresing direction,
(c) its inability to show a minimum input in decresing direction,
(d) none of the above.

Q.21 Dead band of a measuring instrument is -

(a) largest change of measurand to which the instrrument does not respond,
(b) the sum of maximum inputs in both directions to which the instrumnt does not respond,.
(c) the area on the dial of the instreument, in which the instrument does not respond,
(d) none of the above.

Q.22 Drift of a measuring instrument is the variation of output for a given input due to -

(a) speed of input variation, (b) changed conditions,.
(c) load on the output, (d) none of the above.

Q.23 For precise measurement of voltage a VTVM is preferred over an ordinary voltmeter, because -

(a) ordinary voltmeters use inferior components,
(b) a VTVM is a costly hence more accurate devixe,
(c) impedance loading of VTVM is high,
(d) none of the above..

Q.24 In more accurate measurement impedance matching of instruments is adopted, which means -

(a) keeping the impedance of loading circuit high,
(b) keeping the internal impedance of loaded circuit low,
(c) keeping the loading circuit impedance and loaded circuit impedance equal,.
(d) none of the above.

Q.25 The compensating networks in measuring systems are added to improve their dynamic characteristics. For first order instruments, this -

(a) increases the time constant and the static sensitivity,
(b) reduces the static sensitivity and the time constant,.
(c) reduces the time constant and increases the sensitivity,
(d) increases the time constant and reduces the sensitivity.

Q.26 Motion transducers are used to sense (i) displacement, (ii) movement, (iii) speed, (iv) acceleration and may be of mechano-electrical type.

(a) all the above are true, (b) only first three are true,
(c) only first two are true, (d) only first one is true.

Q.27 Basic parameter sensed by a motion trtansducer may be (i) force, (ii) pressure, (iii) torque, (iv) temperature.

(a) all the above are true, (b) only first three are true,
(c) only first two are true, (d) only first one is true.

Q.28 In a motion transducer, the following factors are considered -(i) magnitude of motion, (ii) input-output relationshiip, (iii) the power source required.

(a) all the above are true, (b) only first three are true,
(c) only first two are true, (d) only first one is true.

Q.29 In a potentiometric resistance transducer (i) displacement, (ii) speed, (iii) acceleration can be converted to electrical signal.

(a) all the above are true, (b) only first two are true,
(c) only first one is true, (d) none of the above is true.

Q.30 In an inductive type of transducer (i) displacement, (ii) motion, (iii) speed, (iv) oscillations, (v) acceleration can be converted to electrical signals.

(a) all the above are true, (b) only (i) is false,
(c) only (i) and (iv) are false, (d) only (i) and (v) are false,.

Q.31 In a capacitive type of transducer (i) displacement, (ii) motion, (iii) speed, (iv) oscillations, (v) acceleration can be converted to electrical signals.

(a) all the above are true, (b) only (i) is true,
(c) only (i) and (iv) are false, (d) only (i) and (v) are false,

Q.32 A piezo-electric crystal can be used as a transducer to measure (i) force, (ii) displacement, (iii) motion.

(a) all the above are true, (b) only (i) and (ii) are true,
(c) only (i) is true, (d) none of the above is true.

Q.33 Small displacements can be measured by using the following as transducer (i) resistance strain gauge, (ii) capacitive transducers, (iii) inductive transducers.

(a) all the above are true, (b) only (iii) is false,.
(c) only (ii) is false, (d) only (i) is false.

Q.34 Bonded resistance strain gauges are used as transducers to measure (i) strain, (ii) force, (iii) vibrations, (iv) torque.

(a) all the above are true, (b) only (i) is false,
(c) only (ii) is false, (d) only (iii) is false.

Q.35 An opto-electric analog transducer can be used to meaure (i) motion, (ii) speed, (iii) acceleration.

(a) all the above are true, (b) (i) and (ii) are false,
(c) (ii) and (iii) are false, (d) only (iii) is false.

Q.36 In a moving iron instrument, torque produced is proportional to -

(a) peak value of current, (b) r.m.s. value of current,.
(c) r.m.s value of voltage, (d) none of the above.

Q.37 In a wattmeter, the torque produced is proportional to -

(a) average value of currents in the two coils,
(b) r.m.s. value of currents in the two coils,.
(c) r.m.s. values of current and voltage in the two coils,
(d) none of the above.

Q.38 In a three phase energy meter the neutral wire is essential for measuring energy when the load is -

(a) three phase balanced, (b) three phase unbalanced,.
(c) single phase, (d) none of the above.

Q.39 Basically, potentiometer provides measurement of -
(a) voltage, (b) current,
(c) both voltage or current, (d) none of the above.

Q.40 Which of the following type of instrument will be used to measure ac voltage -
(a) moving iron voltmeter, (b) moving coil voltmeter,
(c) none of the above.

Q.41 In a dynamometer type of insrument, the magnetic field is produced by -
(a) permanent magnets, (b) electromagnets,.
(c) self induction, (d) none of the above.

Q.42 A moving coil type of instrument can not be used for -
(a) ac measurements,. (b) dc measurements,
(c) high voltage measurements, (d) low voltage measurements.

Q.43 Which one type of instruments of the following will have non-uniform calibrations -
(a) moving coil instruments,
(b) moving iron instruments,.
(c) dynamometer type of instruments,
(d) none of the above.

Q.44 Damping in the instruments provides them -
(a) accuracy of measurement, (b) stability of measurement,.
(c) precision of measurement, (d) none of the above.

Q.45 Measuring instruments can be designed with zero damping, although it is not desirable.
(a) true, (b) false,.

Q.46 A hot wire measuring type of instrument can be used for -
(a) ac only, (b) dc only,
(c) both ac and dc, (d) high frequencies only.

Q.47 Creeping phenomenon can be present in -
(a) ameters, (b) voltmeters,
(c) wattmeters, (d) energy meters..

Q.48 Harmonics in a circuit can be found out by the use of the phenomenon of -
(a) induction, (b) resonance,.
(c) electromagnetism, (d) none of the above.

Q.49 The measurements of temperature on celcius and farenheit scale will be same when -
(a) measurement is done with pyrometers,

(b) the tem-perature is -40 degree,.
(c) the temperature is negative,
(d) none of the above.

Q.50 Internal resistance of the following instrument is higher -
(a) ameter,
(b) voltmeter,.
(c) dynamometer type of ameter,

Q.51 Current transformer utilizes the phenomenon of -
(a) induction,. (b) proximity effect,
(c) electrostriction, (d) none of the above.

Q.52 Secondary of a current transformer is always short-circuited to avoid -
(a) primary open-circuiting fault, (b) secondary high voltage,.
(c) primary short circuit, (d) none of the above.

Q.53 A current transformer specification is in the form of a ratio a/b, a and b are respectively -
(a) primary / secondary turns,
(b) secondary / primary turns,
(c) primary / secondary currents,
(d) primary / secondary voltage.

Q.54 Full-scale deflection of a voltmeter is obtained at 10 mA with applied voltage of 500 volts. If its range is desired to be doubled, it has to be added a resistance of -
(a) 50 KOhm in series, (b) 50 KOhm in parallel,
(c) 5 KOhm in series, (d) 5 KOhm in parallel.

Q.55 Full-scale deflection of a moving coil ammeter requires 10 mA at 100 Amps reading. It must be having a -
(a) high resistance in series, (b) high resistance in parallel,
(c) low resistance in parallel, (d) a current transformer.

Q.56 Shunt is used to extend the range of an ameter. Shunt is a -
(a) very low inductance, (b) very low capacitance,
(c) very low resistance, (d) none of the above.

Q.57 If a multimeter range is set to read amperes but is mistakenly connected across the voltage, due to this, -
(a) a very heavy current will flow through the instrument,.
(b) current through the meter will be extremely low,
(c) meter will refuse to work,
(d) none of the above.

Q.58 A multimeter always has a battery in it. This battery is utilized during -

(a) current measurements only,
(b) voltage measurements only,
(c) resistance measurements only,
(d) for all the measurements.

Q.59 An insulation tester is used to measure leakage currrents thhrough the insulation, thereby providing value of insulation resistance. It reads only when connected to -
(a) a circuit with a current source,
(b) a circuit with a voltage source,
(c) a circuit when it is dead,.
(d) none of the above.

Q.60 A Wheatstone bridge is used to measure -
(a) low value of voltage drop, (b) low value of current flow,
(c) resistance values, (d) none of the above.

Q.61 Insulating materials in electrical equipment cause the presence of
(a) only an inductance, (b) only a capacitance,
(c) a resistance and capacitance, (d) a resistance and inductance.

Q.62 Ideally a voltmeter should have -
(a) infinite resistance,
(b) zero resistance,
(c) resistance equal to the source resistance,
(d) none of the above.

Q.63 If a multimeter in resistace measurment setting is connecetd across a healthy capacitor, its needle will -
(a) fast move to low resistance and gradually return to high resistance side,.
(b) slowly move to high resistance side,
(c) not move at all,
(d) none of the above.

Q.64 A 3-phase, 3-wire unbalanced load power can be measured by -
(a) one wattmeter method, (b) two wattmeter method,.
(c) a voltmeter and an ameter, (d) none of the above.

Q.65 The basis of operation of a clectrical reslstance thermometer is -
(a) thermo-electric effect, (b) temperature coefficient,.
(c) thermal expansion, (d) none of the above.

Q.66 A tachogenerator is used for speed measurement as it produces-
(a) current proportional to the speed,

(b) voltage proportional to the speed,.
(c) power proportional to the speed,
(d) none of the above.

Q.67 A stroboscope is used to measure -
(a) speed,. (b) phase difference,
(c) reactive power, (d) nnone of the above.

Q.68 A photoell basically used to -
(a) produce light proportional to some parameter,
(b) sense the presence of light,
(c) always produces voltage proportional to light,
(d) none of the above.

Q.69 A photoconductive cell -
(a) produces light proportional to conductivity,
(b) produces voltage proportional to light,
(c) resistance changes with light,.
(d) none of the above.

Q.70 A photovoltaic cell -
(a) produces light proportional to conductivity,
(b) produces voltage proportional to light,.
(c) resistance changes with light,
(d) none of the above.

Q.71 A hygrometer is an instrument to measure -
(a) viscosity, (b) specific gravity,
(c) humidity, (d) none of the above.

Q.72 Flow measurement is usually done on the basis of measurement of -
(a) temperature difference, (b) voltage difference,
(c) pressure differnce, (d) none of the above.

Q.73 A synchroscope is used to -
(a) measure synchronous speed of alternators,
(b) continuously monitor synchronization,
(c) find synchronization state,.
(d) none of the above.

Q.74 In a power plant, power factor meters are used to find out (i) excitation of alternators, (ii) load division between various machines, (iii) power factor of the load.
(a) all the above are true,
(b) only last two are true,
(c) only (i) and (iii) are true,
(d) only (iii) is true.

Q.75 In a power plant, indicating type of instuments provide (i) operating guidance, (ii) operating supervision, (iii) performance calculations basis.

(a) all the above are true, (b) only first two arer true,
(c) only first one is true, (d) only last two are true.

Q.76 In a power plant, recording type of instuments provide (i) operating guidance, (ii) operating supervision, (iii) performance calculations basis.

(a) all the above are true, (b) only first two arer true,
(c) only first one is true, (d) only last two are true..

Q.77 Integrating type of instruments provide output according to -

(a) current value, (b) variation in value,
(c) accumulated value, (d) none of the above.

Q.78 For temperature measurement, thermocouples of (i) copper-constantan, (ii) chromel-alumal, (iii) platinum-iridium, (iv) platinum-constantan, are used.

(a) only (iv) is false, (b) only (i) is false,
(c) only (ii) is false, (d) only (iii) is false,

Q.79 Voltage produced by an iron-constantan thermocouple at 1000 degree farenheit is -

(a) 22.2 mV, (b) 29.7 mV,.
(c) 15.3 mV, (d) 5.2 mV.

Q.80 Resistance measurement is possible only by flow of current in it.

(a) true, (b) false,

Q.81 A DVM (digital voltmeter) basically reads -

(a) dc voltage, (b) dc current,
(c) ac voltage, (d) ac current.

Q.82 An integrating type digital voltmeter is basically a -

(a) current to voltage drop convertor,
(b) voltage to current flow converter,
(c) voltage to frequency converter,.
(d) current to frequency converter.

Q.83 A cathode ray oscilloscope is basically provides output as a -

(a) voltage operated device, (b) current operated device,
(c) frequency operated device, (d) none of the above.

Q.84 The light visible on a cathode ray tube screen is -

(a) impingement of electrons on the screen,
(b) emission from the screen due to electrons impingement,.
(c) glow due to voltage applied across the deflecting plates,
(d) none of the above.

Q.85 In a CRO a mesurand is displayed agains a timing signal, by connecting by comnnecting measurand and the timing signals respectively at -

(a) horizontal and vertical deflection plates,.
(b) vertical and horizontal deflection plates,
(c) deflection plates and the cathode,
(d) none of the above.

Q.86 To get the phase difference between two waves in the shape of Lissajous diagrams, the two signals are respectively connected to -

(a) the two sets of deflection plates,.
(b) horizontal deflection plate and cathode.
(c) vertical deflection plate and the cathode,
(d) none of the above.

Q.87 A galvenometer is an instrument wherein the measurand -

(a) current-carrying coil moves in a magnetic field,.
(b) voltage produces a field to move a coil,
(c) frequency is utilized for deflection,
(d) none of the above.

Q.88 In reflecting-light type of galvenometers, a mirror is attached to the coil suspension. This type of igalvenometer works well upto a frequency of -

(a) nearly 200 Hz, (b) nearly 10 KHz,.
(c) nearly 100 KHz, (d) none of the above.

Q.89 For data display purposes in digital instruments, LEDs can be used. For the display of a single digit, minimum -

(a) one LED is essential, (b) 3 LEDs are essential,
(c) 7 LEDs are essential, (d) none of the above.

Q.90 The input impedance of a CRO is usually of -

(a) low value, (b) high value,.

Q.91 A CRO can be used to measure dc voltage.

(a) true, (b) false,

Q.92 A CRO can provide simultaneously the magnitude of two independent signals and their phase difference.

(a) true, (b) false,

Q.93 Frequency response of a galvenometer type of recorder, as compared to that of a servo type potentiometric recorder, is -

(a) better, (b) worse,
(c) almost equal.

Q.94 A dc signal can be recorded magnetically on a tape -

(a) and reproduced when desired, (b) but can not be reproduced,.
(c) even recording is not possible.

Q.95 pH value of a solution is a mesure of its acidity and is -
(a) hydrogen ion concentration in grams per cu.meter,
(b) log base 10 of (a) above,

Q.96 Chromatography is a science used for separating and identifying the components in a given mixture, on the basis of -
(a) colour, (b) retention time,.
(c) density, (d) none of the above.

Q.97 Hydrocarbons in the gaseous form can be detected by -
(a) chromatography, (b) spectrometry,.
(c) viscometry, (d) none of the above.

Q.98 Two fluids are flowing in two identical pipes at the same Reynold number. Thedensities of A and B are same, but the pressure drop per unit length is 4 times in the case of A than B. The ratio of viscosities of A and B are -
(a) 1:4, (b) 4:1,
(c) 1:2, (d) 2:1.

Q.99 A vapour-gas mixture becomes saturated at a temperature, which is typical of the vapour and is known as its -
(a) flash point, (b) condensation point,
(c) dew point, (d) none of the above.

Q.100 The hydrogen ion concentration of a solution is 2.5×10^{-11} g/liter. The solution is -
(a) neutral, (b) acidic,
(c) alkaline, (d) can't be determined.

Q.101 The ECG traced by a medical expert, represents -
(a) the record of intensity of heart-beats,
(b) the record of electric pulses in the cardiac muscle,
(c) the curve of depolarization and repolarization of myocardium,.
(d) none of the above.

Q.102 The most common carrier gas employed in gas chromatography is -
(a) methane, (b) helium,.
(c) oxygen, (d) none of the above.

Q.103 Non-dispersive infra-red spectrometry is suitable for detecting -
(a) oxygen, (b) methane,.
(c) nitrogen, (d) none of the above.

Q.104 A chromatograph is used for measuring -
(a) flow rate of a gas, (b) the composition of a gas,.
(c) temperature of a gas, (d) none of the above.

ABSTRACT

8.1. DEFINITIONS

Instrument : A device used for determining the value or magnitude of a property or a variable of a thing, system or concept.

Accuracy : It is a statement of the limits which bound the departure of the measured value from the true value of a quantity under measurement. It is a characteristic of measurement system.

Precision : It is a measure of the spread of repeated determination of a particular quantity. It is the characteristic of the measuring instrument or method.

Resolution : It is the smallest change in the measured quantity that can be detected by the instrument with reasonable certainty.

Threshold : It is the minimum value of the measured quantity to which the instrument will provide response. Thus it is the resolution at minimum point of scale.

Static sensitivity : It is the ratio of the instrument response to the magnitude of the quantity under measurement or

$$k_s = \frac{\text{Change in output signal of instrument}}{\text{Change in output signal}} \text{ at input value}$$

Deflection factor : It is the reciprocal of sensitivity.

Hysterisis : This the property of the instrument of providing different measurements for the same quantity when the quantity is gradually increased from a lower value and when it is gradually decreased from a higher value.

Dead band : This is the largest change in instrument input of zero level, in both directions to which the instrument does not respond.

8.2. THEORY OF ERRORS

All measurement results are stated as a numerical value, a unit and a level of uncertainty. The uncertainty is caused by the errors, which can not be fully eliminated.

The prime sources of errors are :

(i) Noise signals,
(ii) Response delay,
(iii) Design limitations,
(iv) Extraction of measured quality by the measurement system,
(v) Transmission losses,
(vi) Physico-chemical deterioration,
(vii) Ambient influences.

Errors may be classified as :

(a) Systematic errors, and
(b) Random errors.

Data evaluation

Extraction of the right results from the data full of uncertainties is a subject of statistical analysis. Some hints are provided.

(i) Standard deviation (σ)

$$\sigma = \sqrt{\frac{\Sigma d_m^{\ 2}}{n}}$$

where d_m = individual deviation of the results from mean value.
n = number of observations.

(ii) Probable error (p_e)

$p_e = \pm 0.6745 \times \sigma$

(iii) Confidence interval of a set of values is the range of deviation from the mean within which a certain fraction of the observed values are expected to lie. Standard statistical tables are used for this analysis.

(iv) Test for systematic error : If σ_A and σ_b are the standard deviation values for two sets of measurements and A and B are the average of these two sets, then a systematic error is present if :

$A - B > 2(\sigma_A - \sigma_B)$

8.3. MEASUREMENT SYSTEMS

The selection of a measurement system is done with a view to maximise the payoff ratio of the system, which is defined as :

$$\frac{\text{Value of information obtainable from the system}}{\text{Cost of the system}}$$

A measurement system performs the following functions :

(i) Sensing a quantity and converting it to a useful signal.
(ii) Conditioning of the signal, like amplification.
(iii) Signal transmission to the measuring of displaying device.
(iv) Display of results.

8.4. MEASUREMENT OF ELECTRICAL QUANTITIES

The table below shows the types of instruments, which may be used for various electrical measurements.

Type of instrument	*Current*		*Voltage*		*Power*		*Energy*	
	AC	*DC*	*AC*	*DC*	*AC*	*DC*	*AC*	*DC*
(a) Moving-iron	√	√	√	√	-	-	-	-
(b) Moving-coil								
(i) Permanent magnet	X	√	X	√	-	-	-	-
(ii) Dynamo meter	√	√	√	√	√	√	-	-
(c) Induction type	√	X	√	X	√	-	√	-
(d) Hot wire	√	√	√	√	-	-	-	-
(e) Electrostatic	X	X	√	√	-	-	-	-

8.4.1. Analog instruments

A. Moving-iron instruments

The arrangement consists of a solenoid coil producing a magnetic field which induces magnetization in the two iron vanes placed in the field. One vane is fixed while the other is moving. When no current is in the solenoid coil, both are close to each other, held by springs.

When a current passes in the coil, the two vanes repel each other resulting in a displacement of the moving vane, which carries the indicating needle with it.

The torque experienced by the moving vane is given by :

$$T = \frac{1}{2} I^2 \frac{dL}{d\theta}$$

where I = Operating current
L = Inductance of the solenoid coil
θ = angular deflection of the moving vane.

B. Moving-coil instrument

It consists of a moving coil in a magnetic field produced by a permanent magnet. The current to be measured is passed through the moving coil. When there is no current in the coil, it is held by the springs in a plane parallel to the magnetic field. The final steady deflection is given by :

$$B = \frac{G}{k_o} .I$$

where G = NBlD
where N = Number of turns in the coil
B = Flux density
l = Length of coil perpendicular to the direction of field
d = length of the coil side parallel to the field
k_o = Spring constant
I = Current to be measured.

C. Dynamometer type instrument

This arrangement consists of a fixed coil to produce a magnetic field and a pivoted moving coil, both coil magnetically coupled together.

The fundamental torque produced by the arrangement is given by :

$$T = I_1.I_2 \frac{dM}{d\theta}$$

where I_1 and I_2 are the two coil currents
M is the coefficient of mutual inductance between the two coil
θ is the angular deflection

since M is maximum when $\theta = 0$ and minimum when $\theta = 90$

$M = M_o \cos\theta$

Hence $T = I_1 I_2 (-\sin Q)$

$= I_1 I_2 \sin\theta$

For measurement of current and voltage the two coils are connected in series. Hence $I_1 = I_2 = I$ therefore :

$T = I^2 \sin\theta$

The most important application of dynamometer type of instruments is in wattmeters, where one coil is connected to the voltage and the other to the current. The two coils are specially designed for such purposes.

For power measurement in 3-phase circuits either two wattmeter method or 3-wattmeter method may be used.

In three-wattmeter method, one wattmeter is connected across each phase. The total power transfer is given by the sum of the three wattmeter readings.

The two-wattmeter method, the measurements are subject to the load power factor, such as :

(i) when power factor is unity.

$W_1 = W_2$

The total power transfer = $W_1 + W_2 = 2W_1$

(ii) When powerfactor is 0.5,

$W_1 = 0$

The total power transfer = W_2

(iii) When power factor is between 0 and 0.5,

One watt meter reading less will give negative measurement. Total power transfer = $W_1 - W_2$

(iv) When power factor is zero :

$W_1 = - W_2$

Total power transfer = $W_1 - W_2 = 0$

(v) When the load is balanced, the power factor angle is given by :

$$\tan\theta = \frac{\sqrt{3}\,(W_1 - W_2)}{(W_1 + W_2)}$$

8.4.2. Galvanometers

Galvanometers are vary sensitive instruments used to detect or measure the small currents or voltages. These are mostly used in bridge circuits for null detection.

(i) D'Arsonval galvanometer : Its basic mechanisms is like that of a moving coil instrument, but more sensitive. The indication of deflection is provided through reflection of a fixed light beam, by a mirror attached to the deflecting mechanism. The deflection of the reflected light is measured at a distance of 1m.

(ii) Ballistic galvanometer : The D'Arsoval mechanism is fed by the output of a search coil. Presence of a magnetic flux in the vicinity of the search coil induces a voltage in it which gives deflection in the galvanometer proportional to the induced emf.

8.4.3. Potentiometers

A potentiometer measures an unknown voltage by comparing it with a known voltage dropped in a calibrated resistance assembly. When the two voltages are equal, the deflection in the galvanometer, which is connected to read the difference, is zero. Such a method is more accurate than analog instruments.

8.4.4. Bridge methods

Bridge methods provide accurate measurements for resistances, inductances and capacitances. The basic principle of a bridge is balancing of voltages at the mid points of two series circuits by varying the circuit impedence or resistances. The mid points are connected to a galvanometer to detect the balance. When the balance is reached the ratios of the impedances of the two circuits before and after the mid points are same. If three values of the bridge so formed are known, the value of the fourth arm is easy to calculate.

8.5. INSTRUMENTATION

Measurement of a nonelectrical quantity through electrical measurements is usually considered as instrumentation. Another notable feature of instrumentation is that it can provide signal to an automatic control system to regulate the quantity.

An instrumentation system consists of a transducer, which provides electrical signal as a function of the measured quantity, some signal conditioning device and a display system. The heart or key to instrumentation is the transducer. A large variety of transducers to suit various types of inputs and outputs, are being used by instrumentation engineers. A summary of some such transducers is provided below :

8.5.1. Transducers

Measured	*Transducer*	*Function*
Processure, force or displacement, stress torque	1. Spring loaded potentiometer	The slider of the potentiometer slides on application of pressure varying the resistance.
	2. Capacitance transducer	Distance between the electrodes of capacitor is varied by application of force or displacement.

(Contd.)

Measured	*Transducer*	*Function*
	3. Inductance transducer	Measur and varies the coupling of a coil with magnetic circuit changing inductance of coil.
	4. Reluctance transducer	Measur and changes the reluctance of a core by moving an external iron piece closer or away from it.
	5. Differential transformer transducer	The measur and varies the coil position or the core position of a differential transformer's two secondaries varying the output.
	6. Eddy-current transducer	Loading of a coil is varied by moving on eddy-current plate near the core
	7. Magnetostriction transducer	Measur and varies the magnetic properties of the core by magnetostriction.
Gas flow	1. Pirani gage	The flow of a gas cools down a current-carrying resistor changing its resistance
	2. Hot-wire meter	-do-
Temperature, Radiant heat, hot blast	1. Resistance thermometer	The resistance of a pure metal wire, with positive temperature coefficient of resistance, varies with temperature variation.
	2. Thermistor	Resistance of metal oxide or ceramic materials having negative temperature coeft of resistance, varies with temperature variation.
	3. Thermocouple	The emf generated at a junction of two dissmiliar metals varies with temperature variation.
	4. Thormopile	The emf generated at the junction of two dissimilar semiconductors varies with temperature variation.
Humidity	Resistance hydrometer	Resistance of a hygroscopic conductive strip varies with moisture content.
Light, an object obstructing the light	1. Photo-conductive cell	Distance of a photosensitive semiconductor cell varies with incidence/obstruction of light.
	2. Photo-emissive cell	The incident light causes the emission of electrons from the cell surface.
	3. Photo-multiplier tubc	The incident light causes the secondary electron emission from photosensitive cathode.
	4. Photoroltaic cell	The incident light stimulates a semiconductor junction to generate a voltage.

(Contd.)

Measured	Transducer	Function
Speech, music, noise, sound	1. Capacitive microphone	The sound vibrations vary the capacitance between a fixed plate and a flexible diaphragm.
	2. Piezoelectric pickup	Piezoelectric crystal such as quartz generate a voltage on the application of a force or vibration.
	3. Carbon microphone	The sound vibrations vary the resistance of carbon powder filled in a cell.
Liquid-level	Capacitive gage	The liquid level when reaches between the two plates of a capacitor, its capacitance changes.
Thickness	1. Capacitive meter	The displacement of capacitor plate in accordance with the measured thickness varies the capacitance.
	2. Eddy-current gage	The proximity of an eddy-current plate with an inductive circuit flux, in accordance with the thickness varies the loading of the coil.
Vibrations	1. Reluctance pickup	The reluctance of a magnetic circuit varies with variation in core due to vibrations.
	2. Moving coil inducer	Movement of a coil in a magnetic field induces a voltage in it.
Magnetic flux, current	Hall-effect transducer	A potential difference is generated in a semiconductor plate when a magnetic flux interacts with a current.
Radioactivity	Ionization chamber	Radioactive radiations induce gas ionizations causing change in conductivity.
Velocity	Moving-coil generator	Rotation of a coil in a magnetic field induces a voltage in the coil.

8.5.2. Signal conditioners

Signals provided by a transducer may have undesirable features such as noise, disproportionality with the input, or a form different from the desired form. To bring these signal to the usable form for the display device is called signal conditioning. The following techniques are used for this purpose :

1. Amplification

Amplifiers make a weak signal strong and perceivable by the display device. Electronic power amplifiers are very popular.

2. Compensators

Compensators are used to improve the frequency response and impedance loading between the transducer and the display device. R-C circuits in various combinations perform this function.

3. Differentiators/Integrators

The transducer output may have, let us say, a velocity and acceleration signals together. To choose the one and suppress the other is done by differentiation or an integrator circuit. These circuits use amplifiers with resistance capacitance combinations to perform the desired function.

4. Filters

Filters suppress the noise from the signal. It may be a low-pass, high-pass, a band-pass or band suppressor filter depending upon the situations.

5. ADC/DAC circuits

For conversion of analog signal to digital or vice-versa analog-to-digital converters or digital-to-analog converters are used. Potentiometric converters are the simplest to perform.

9

Electromagnetic Waves and Telecommunication

Q.1 The radiowaves have been classified into 8 groups starting from very-low frequency to Extra-high frequency, with ranges having upper limit 10 times the lower limit, starting from -

(a) VLF as below 30 kHz, (b) VLF as above 30 kHz,
(c) VLF as below 300 kHz, (d) none of the above.

Q.2 In communication, essential feature of a message are (i) its unpredictability, (ii) its repeatability, (iii) its transmittability.

(a) all the above are true, (b) only first two are true,
(c) only last two are true, (d) none of the above is true.

Q.3 In telecommunication, a transducer converts the physical quantity of the message to a transmissible signal, mostly -

(a) electrical wave, (b) sound wave,
(c) light wave, (d) none of the above.

Q.4 The wavelength of an electromagnetic wave of frequency f, is given by the expression -

(a) $3 \times 10^8 \times f$ meters, (b) $3/10^8 \times f$ meters,
(c) $3 \times 10^8 /f$ meters, (d) none of the above.

Q.5 The minimum frequency of an electromagnetic wave, in the VLF range is -

(a) 3 kHz, (b) 16 kHz,
(c) 30 kHz, (d) none of the above.

Q.6 A channel is a means of communication from transmitter to receiver and thus is -

(a) unidirectional, (b) bidirectional,
(c) any of the above, (d) none of the above.

Q.7 A single channel, may be used to transmit multiple signals signals, simultaneously, with each message -
(a) in a different frequency band, (b) at a different time-slice,
(c) in different amplitude, (d) none of the above.

Q.8 According to international agreement, a single cable may be used to transmit -
(a) 270 channels, (b) 1,200 channels,
(c) 2,700 channels, (d) none of the above.

Q.9 Electrical energy is transmitted in the form of electromagnetic waves.
(a) always, (b) never,
(c) sometimes. (d) mostly.

Q.10 When electrical waves can travel in space due to radiattion, they are called -
(a) electromagnetic waves, (b) radio waves,
(c) microwaves, (d) none of the above.

Q.11 When the electromagnetic waves are carrieded by transmission lines, the mode of propagation is called -
(a) transmission mode, (b) principal mode,
(c) low-frequency mode, (d) none of the above.

Q.12 A wave is called a transverse electromagnetic (TEM) wave, when -
(a) electric and magnetic fields are at right-angles,
(b) electric and magnetic fields move in opposite directions,
(c) magnetic fields are absent,
(d) none of the above.

Q.13 Any signal pressed into a finite transmission line, produces two voltage waves, travelling -
(a) in opposite directions, (b) on the two wires of line,
(c) at right angles to each other, (d) none of the above.

Q.14 The characteristic impedance of a transmission line, is its impedance at the input terminalse, when the line is treated to be of -
(a) infinite length, (b) finite length,
(c) short length, (d) none of the above.

Q.15 Voltage and current phasors on a transmission line are continuously diminished with length with their phase angles -
(a) remaining fixed, (b) alternating,
(c) continuously increasing, (d) none of the above.

Q.16 For an electromagnetic wave, with a travel of distance equal to its wavelength, the phase angle of its phasor changes by -

(a) π radians, (b) 2 π radians,
(c) $\pi/2$ radians, (d) none of the above.

Q.17 Phase velocity of a wave denotes the location of a particular phase in space ans given by -
(a) wavelength x frequency, (b) wavelength/frequency,
(c) frequency/wavelength, (d) none of the above.

Q.18 In an infinite transmission line there is no loss of energy due to reflection. Hence a finite line is made to behave like an infinite line, by terminating it with an impedance, that is -
(a) equal to the characteristic impedance of the line,
(b) more than the characteristic impedance of the line,
(c) less than the communication impedance of the line,
(d) none of the above.

Q.19 Characteristic impedance of a line may be defined as its impedance at the input terminals, when -
(a) the line length is infinite,
(b) the character of the input and output are same,
(c) the phase shift introduced by the line is zero,
(d) none of the above.

Q.20 If the output end of a transmission line is open-circuited, the magnetic field collapses, giving rise to -
(a) voltage collapsing, (b) voltage doubling,
(c) infinite impedancing, (d) none of the above.

Q.21 If the output end of a transmission line is short-cicuited, the electic field collapses, giving rise to -
(a) current collapsing, (b) current doubling,
(c) zero impedancing, (d) none of the above.

Q.22 If two transmission lines with characteristic impedances respectively Z_o and Z_r are cascaded, the voltage reflection coefficient is given by -
(a) $(Z_r - Z_o)/(Z_o + Z_r)$, (b) $2 Z_r /(Z_o + Z_r)$,
(c) $2 Z_o /(Z_o + Z_r)$, (d) $(Z_o - Z_r)/(Z_o + Z_r)$.

Q.23 If two transmission lines with characteristic impedances respectively Z_o and Z_r are cascaded, the voltage transmission coefficient is given by -
(a) $(Z_r - Z_o)/(Z_o + Z_r)$, (b) $2 Z_r /(Z_o + Z_r)$,
(c) $2 Z_o /(Z_o + Z_r)$, (d) $(Z_o - Z_r)/(Z_o + Z_r)$.

Q.24 If two transmission lines with characteristic impedances respectively Z_o and Z_r are cascaded, the current reflection coefficient is given by -

(a) $(Z_r - Z_o)/(Z_o + Z_r)$, (b) $2 Z_r /(Z_o + Z_r)$,
(c) $2 Z_o /(Z_o + Z_r)$, (d) $(Z_o - Z_r)/(Z_o + Z_o)$.

Q.25 If two transmission lines with characteristic impedances respectively Z_o and Z_r are cascaded, the current transmission coefficient is given by -

(a) $(Z_r - Z_o)/(Z_o + Z_r)$, (b) $2 Z_r /(Z_o + Z_r)$,
(c) $2 Z_o /(Z_o + Z_r)$, (d) $(Z_o - Z_r)/(Z_o + Z_r)$.

Q.26 If R, G, L and C are respectively the series resistance, shunt conductance, series inductance and shunt capacitance of a line, the distortionless condition of the line, at all frequencies, is -

(a) LC = RG, (b) LG = RC,
(c) LR = GC, (d) none of the above.

Q.27 Two parallel running transmission lines can be used for transmission of three speech signals by using -

(a) one conductor from one line and one conductor from the other,
(b) phantom working,
(c) reciprocal working,
(d) none of the above.

Q.28 An ordinary transmission line provides different attenuation and transmission delays to different frequencies of a message, hence the message may get distorted beyond recognition. This is eliminated by using -

(a) balancers, (b) filters,
(c) equalizers, (d) none of the above.

Q.29 Two impedances are said to be inverse of each other, if the geometric-mean of the two is -

(a) positive, (b) negative,
(c) a real number, (d) none of the above.

Q.30 An equalizer network for a communication transmission line is composed of -

(a) complimentary impedances, (b) dual impedances,
(c) inverse impedances, (d) none of the above.

Q.31 If amplitude of an amplitude-modulated wave varies between (A + B) and (A - B), its degree or depth of modulation is -

(a) 2B, (b) B/A,
(c) AB, (d) none of the above.

Q.32 An amplitude-modulated carrier is said to be fully modulated, when its amplitude varies between -

(a) zero and twice the unmodulated amplitude,
(b) zero and unmodulated amplitude,

(c) unmodulated amplitude and twice of it,
(d) none of the above.

Q.33 In amplitude modulation. the modulation percentage is, usually, kept -
(a) 100 percent,
(b) slightly less than 100 percent,
(c) 50 percent,
(d) none of the above.

Q.34 As compared to the bandwidth of the original signal, the bandwidth occupied by the amplitude-modulated wave is -
(a) equal, (b) half,
(c) twice, (d) none of the above.

Q.35 In amplitude modulation, if power of the modulated carrier is P, the power of the modulating signal, should be -
(a) P/2, (b) P/3,
(c) 2P/3, (d) none of the above.

Q.36 In frequency modulaation, the amplitude of the carrier wave is kept constant and its frequency is modified, so that the modulated-signal frequency is such that -
(a) it is proportional to signal amplitude,
(b) its deviation is proportinal to signal amplitude,
(c) its phase angle is proportional to signal amplitude,
(d) none of the above.

Q.37 In frequency-modulation, ratio of deviation in carrier wave frequency and signal frequency is known as -
(a) deviation ratio, (b) modulation index,
(c) modulation ratio, (c) none of the above.

Q.38 In phase modulation, the phase of the carrier wave is advanced or retarded by an angle proportional to -
(a) signal frequency, (b) signal phase-angle,
(c) signal intantaneous amplitude, (d) none of the above.

Q.39 In pulse modulation, the carrier wave is modified by the use of a pulsed wave. The signal being carried in some form by -
(a) carrier wave, (b) pulse wave,

Q.40 In a communication line, the reflected wave is avoided by -
(a) connecting a sink capacitor at receiving end,
(b) adding unidirectional impedanc elements in the line,
(c) terminating the line by its characteristic impedance,
(d) none of the above.

Q.41 Amplitude modulation can be done by using the circuit known as (i) Carson modulator, (ii) Cowan modulator, (iii) Ring modulator.

(a) all the above are true, (b) only first two are true,
(c) only last two are true, (c) none of the above is true.

Q.42 In Carson circuit of amplitude modulation, carrier wave and modulating signal are superimposed through the base circuit of two transistors or vacuum tubes connected in -

(a) cascade, (b) parallel,
(c) push-pull, (d) none of the above.

Q.43 Ring and Cowan modulator circuits can be used for -

(a) modulation only,
(b) demodulation only,
(c) both modulation and demodulation,
(d) none of the above.

Q.44 Reactance valve modulator is applicable to -

(a) amplitude modulation, (b) frequency modulation,
(c) phase modulation, (d) none of the above.

Q.45 In radio transmitter, a wave is produced by -

(a) only a train of voltage pulses,
(b) each uprise of exciting voltage,
(c) each periodic pulse of exciting voltage,
(d) none of the above.

Q.46 In an electromagnetic wave, the three vectors of electrostatic field, magnetostatic field and the wave propagation are -

(a) always mutually perpendicular,
(b) may be mutually perpendicular,
(c) can't be mutually perpendicular.

Q.47 The wavelength of a radiowave is -

(a) proportional to its frequency,
(b) inversely proportional to its frequency,
(c) equal to that of light,
(d) none of the above.

Q.48 For a radio wave, electric and magnetic components are in phase with other, are at right-angles in space, and the ratio of their magnitudes is known as impedance of free space, with a value of -

(a) 120 ohms, (b) 220 ohms,
(c) 377 ohms, (d) none of the above.

Q.49 The magnetic flux of a radio wave, has a circular path centered around the souce conductor and the flux density -

(a) is steady,
(b) decreases with radial distance from the source,
(c) increases with radial distance from the source,
(d) none of the above.

Q.50 The electric field of a radio wave starts between the ends of the source conductor, and wrt. the magnetic field, it is -

(a) orthogonal, (b) homoplanar,
(c) hexagonal, (d) none of the above.

Q.51 The rate of decay of electric field, as compared to the decay of potential that produces it, is -

(a) same, (b) lower,
(c) higher.

Q.52 For a radio wave, if E is the electric field, in volts/meter, and H is the magnetic field in amperes/meter, the power density P, carried by each sq.m. of the wavefront, known as Poynting vector, is given by -

(a) E/H watts/sq.m., (b) E.H watts/sq.m.,
(c) H/E watts/sq.m., (d) none of the above.

Q.53 For a radio wave produced by a current element of length L meters, which is very short as compared to the wavelength W_l, the radiation resistance is given by -

(a) $80\ \pi^2\ (L/W_l)^2$ ohms, (b) $80/\pi^2\ (L/W_l)^2$ ohms,
(c) $80\ \pi\ (L/W_l)^2$ ohms, (d) none of the above.

Q.54 For a radio wave, the power radiated by a half-wavelength dipole is independent of the frequency and for a current of I amperes, is given by -

(a) $80\ I^2$, (b) 80 I,
(c) 80/I, (d) none of the above.

Q.55 To be precise, the radiation resistance for a half-wavelength dipole is nearly -

(a) 80 ohms, (b) 73 ohms,
(c) 36 ohms, (d) none of the above.

Q.56 For a vertical aerial for low-frequequency radio waves (wavelength- W_l), impedance depends upon its length, it is capacitive or inductive depends upon whether its length (L) is, respectively -

(a) $L < W_l/4$ or $W_l/4 < L < W_l/2$,
(b) $L > W_l/2$ or $W_l/4 < L < W_l/2$,

(c) $W_l/2 > L > W_l/4$ or $L < W_l/4$,
(d) none of the above.

Q.57 For medium frequency broadcasting aerial vertical height is kept -
(a) less than quarter-wavelength, (b) less than half-wavelength,
(c) more than half-wavelength, (d) none of the above.

Q.58 For high frequency radio transmission, the aerial is chosen to be -
(a) half-wavelength dipole, (b) quarter-wavelength dipole,
(c) full-wavelength dipole, (d) none of the above.

Q.59 For a local public radio broadcasting system with range not exceeding 150 km, the transmitter capacity is kept nearly -
(a) 50 kw, (b) 100 kw,
(c) 200 kw, (d) 400 kw.

Q.60 For a local public radio broadcasting system with range not exceeding 150 km, the radiation field strength range is nearly -
(a) 10 mirovolt/m to 10 mv/m,
(b) 100 microvolt/m to 100 mv/m,
(c) 1 mv/m to 1 v/m,
(d) none of the above.

Q.61 In microwave radio communication, at repeater stations, signal is (i) amplified, (ii) demodulated, (iii) carrier frequency changed, (iv) retransmitted
(a) all the above are true, (b) only (i) is false,
(c) only (ii) is false, (d) only (iii) is false,

Q.62 The optimum frequency for satelite communication system is -
(a) 2 to 8 GHz, (b) 5 to 12 GHz,
(c) 8 to 16 GHz, (d) none of the above.

Q.63 For coverage of total earth throgh satelite communication with three equally-spaced satelites, the satelites are placed at a height of -
(a) 24,000 km, (b) 36,000 km,
(c) 48,000 km, (d) none of the above.

Q.64 Electromagnetic waves are those wave which have a frequency range of -
(a) 30 KHz to 416,000 GHz, (b) 3 MHz to 300 GHz,
(c) 300 MHz to 3000 GHz, (d) none of the above.

Q.65 Microwaves are those wave which have a frequency range of -
(a) 30 KHz to 416,000 GHz, (b) 3 MHz to 300 GHz,

(c) 300 MHz to 3000 GHz, (d) none of the above.

Q.66 Radio-frequency waves are those wave which have a frequency range of -

(a) 30 KHz to 416,000 GHz, (b) 3 MHz to 300 GHz,
(c) 300 MHz to 3000 GHz, (d) none of the above.

Q.67 Microwaves above 3,000 GHz are called -

(a) infrared rays, (b) ultra microwaves,
(c) r.f waves, (d) none of the above.

Q.68 As the frequency of a wave increases, the sizes of antennas and waveguides required to handle the same power -

(a) increasese, (b) decreases,
(c) is not afffected,

Q.69 For point communication, radar and guiodance systems, the following type of wave is used -

(a) radio frequency, (b) microwave,
(c) any e.m. wave, (d) none of the above.

Q.70 For low power point to point communications, the folowing frequency range is used -

(a) 1.12 - 2.60 GHz (b) 2.6 - 18.0 GHz,
(c) 30 - 300 GHz, (d) none of the above.

Q.71 For high power long distance communications, the following frequency range is made use of-

(a) 1.12 - 2.60 GHz. (b) 2.6 - 18.0 GHz,
(c) 30 - 300 GHz, (d) none of the above.

Q.72 Long range radars use the following range of microwave frequencies -

(a) 0.5 - 1 GHz, (b) 1.12 - 2.6 GHz,
(c) 2.6 - 18.0 GHz, (d) none of the above.

Q.73 The guidance of space missiles is carried out by the following range of microwaves -

(a) upto 12.4 GHz, (b) above 30 GHz,
(c) 12.4 - 30 GHz, (d) none of the above.

Q.74 If Z and Y are the series impedance per unit length and parallel admittance per unit length of a transmission line, its propagation constant and the characteristic impedance are respectively given by -

(a) $\sqrt{(ZY)}$ and $\sqrt{(Z/Y)}$, (b) $\sqrt{(Z/Y)}$ and $\sqrt{(ZY)}$,
(c) ZY and Z/Y, (d) Z/Y and ZY.

Q.75 The energy radiated by an electric dipole is maximum along -

(a) its axis, (b) its equatorial plane,
(c) polar planes, (d) none of the above.

Q.76 The total energy radiated by a dipole of dipole-moment p at frequency f, is proportional to -

(a) $p^2.f^2$, (b) p^2/ f^2,
(c) $p.f^2$, (d) none of the above..

Q.77 If s be the distance between the charges of a dipole and f theR frequency of wave, the raddiation resistance is given by -

(a) 20 $(sf)^2$, (b) 20 sf,
(c) 20 s/f, (d) none of the above.

Q.78 A half-wave antenna is a straight conductor fed at the centre-breakage, thus each piece is of -

(a) half-wavelength, (b) quarter-wavelength,

Q.79 A half-wave antenna, often used in radio broadcast, is a vertical mast of quarter-wavelength placed on the ground. The ground provides the other pole as -

(a) a flat surface, (b) an image of the mast,
(c) a crossed line, (d) none of the above.

Q.80 Two electric dipoles of opposite polarity arranged in line to give three charges +Q, -2Q and +Q, is called -

(a) linear electric tripole, (b) linear electric quadrupole,
(c) linear electric dipole, (d) none of the above.

Q.81 Both electric and magnetic dipoles can be used as receiving antennas.

(a) wholly true, (b) wholly false,
(c) partially true.

Q.82 For a short electric dipole, voltage on a high resistance load is equal to -

(a) tangential electric field X dipole length,
(b) tangential electric field / dipole length,
(c) tangential electric field x wave-length,
(d) none of the above.

Q.83 In case of two antennas, one working as a source and the other as a receiver, according to reciprocity theorem, if their roles are reversed, and other thlngs remaining the same -

(a) receiver-current/source-voltage will remain same,
(b) source-current/receiver-voltage will remain same,
(c) source-voltage/receiver-voltage will remain same,
(d) none of the above.

Q.84 In electrostatics, Coulomb's law states that the force between two point charges Q1 and Q2 in coulombs, located at a distance of r meters apart, the force between them, acting along the line joining them, is proportional to -

(a) $Q1.Q2/r^2$, (b) Q1.Q2/r,
(c) Q1.Q2.r, (d) none of the above.

Q.85 The force working between two point charges, is proportional to -

(a) the permitivity of the medium,
(b) the inverse of permitivity of the medium,
(c) the permeability of the medium,
(d) none of the above.

Q.86 An electric dipole means -

(a) two chages separated by a distance,
(b) two equal charges separated by a distance,
(c) two equal and opposite charges separated by a distances,
(d) none of the above.

Q.87 For two charges +Q and -Q, separated by a distance s, the dipole moment, directed from positive to negative charge, is given by -

(a) Q^2/s, (b) Q/s^2,
(c) Qs, (d) none of the above.

Q.88 Two straight parallel wires, with a distance d between them, carry currents I_1 and I_2 in the same direction, the force of interaction between the conductors, is -

(a) attractive, proportional to $I_1.I_2/d^2$,
(b) repulsive, proportional to $I_1.I_2/d^2$,
(c) attractive, proportional to $I_1.I_2/d$,
(d) repulsive, proportional to $I_1.I_2/d$.

Q.89 The magnetic dipole moment, of a loop of area S and carrying a current I is -

(a) I/S, (b) IS,
(c) IS^2, (d) none of the above.

Q.90 A medium is called isotropic, if its properties are -

(a) same at all points, (b) same in all directions,
(c) linearly variable, (d) none of the above.

Q.91 Crystalline media are usually -

(a) isotropic, (b) homogeneous,
(c) anisotropic, (d) none of the above.

Q.92 The three conditions D = εE, B = μH, and J = εE of electromagnetism, are applicable to -
(a) all media, (b) isotropic media,
(c) linear and isotropic media, (d) none of the above.

Q.93 In electromagnetic wave theory, dispersion means -
(a) scattering of waves,
(b) dependence of wave velocity on frequency,
(c) dispersal of waves,
(d) none of the above.

Q.94 The four fundamental equations of electromagnetism, are known as -
(a) Wave equations, (b) Claussius-Mossotti equations,
(c) Maxwell's equations, (d) none of the above.

Q.95 The four Maxwell's equations are applicable to electromagnetic phenomena in -
(a) homogeneous media only, (b) linear media only,
(c) isotropic media only, (d) all static media,

Q.96 Velocity of electromagnetic waves in free space is given by -
(a) $\mu_o\ \varepsilon_o$, (b) $(\mu_o\ \varepsilon_o)^2$,
(c) $1/\mu_o\ \varepsilon_o$, (d) $1/\sqrt{(\mu_o\ \varepsilon_o)}$,

Q.97 For an electromagnetic wave of E and H fields, as usual, direction of its Poynting vector is -
(a) direction of wave propagation,
(b) opposite the direction of wave propagation,
(c) null,
(d) none of the above.

Q.98 Direction of propagation of an electromagnetic wave in free space is -
(a) transverse to E and H, (b) opposite of (a),
(c) perpendicular of (a), (d) none of the above.

Q.99 The E and H vector energy densities, of an electromagnetic wave travelling in free space, are -
(a) in phase and equal, (b) out of phase and equal,
(c) in phase and unequal, (d) none of the above.

Q.100 For an electromagnetic wave travellig in an imaginary free space cylinder, if the amplitude remains constant, the energy flow out will be -
(a) constant, (b) zero,
(c) infinity, (d) none of the above.

Q.101 For an electromagnetic wave travelling in a homogeneous, isotropic and stationary media, the phases of its E and H vector -
(a) will be same, (b) may be different,
(c) will not be same, (d) none of the above.

Q.102 For an electromagnetic wave travelling in a nonconductor medium, the E and H vectors and their energy densities, will be respectively -
(a) in phase and equal, (b) out of phase and equal,
(c) in phase and unequal, (d) none of the above.

Q.103 For an electromagnetic wave travelling in conducting medium, the E and H vectors and their energy densities, will be respectively -
(a) in phase and equal, (b) out of phase and equal,
(c) in phase and unequal, (d) out of phase and unequal.

Q.104 For an electromagnetic wave travelling in highly conducting medium, the propagation of wave will be -
(a) with negligible attenuation, (b) upto the skin depth only,
(c) fully reflected, (d) none of the above.

Q.105 The electromagnetic wave propagation through an ionized gas depends upon wave frequency, whether it is less or more than the plasma frequency of the gas. The propagation at the low and high frequencies are respectively -
(a) zero and as that in free space,
(b) as that in free space and zero,
(c) as in nonconducting and conducting media,
(d) none of the above.

Q.106 For the propagation of an electromagnetic wave from one medium of refraction index n_1 to another with refraction index n_2, the Snell's law provides relationship between the angles of incidence (A_1) and refraction (A_2), as follows ;
(a) $n_1 \times \text{Sin } A_1 = n_2 \times \text{Sin } A_2$,
(b) $n_1 \times \text{Sin } A_2 = n_2 \times \text{Sin } A_1$,
(c) $\text{Sin } A_1 / \text{Sin } A_2 = n_1 \times n_2$,
(d) none of the above.

Q.107 For the propagation of an electromagnetic wave from one medium to another, the relation of amplitudes and planes of the incident, reflected and refracted waves is provided by -
(a) Fresnel's equations, (b) Snell's law,
(c) Maxwell's equations, (d) none of the above.

Q.108 For the propagation of an electromagnetic wave from one medium to another, if coefficients of reflection (R) and refraction (T) give the fractions of the incident power, the following is true -

(a) R x T = 1, (b) R + T = 1,
(c) R / T = constant, (d) none of the above.

Q.109 For the propagation of an electromagnetic wave from one dielectric medium to conducting medium, the coefficient of reflection is -

(a) nearly zero, (b) nearly one,
(c) nearly 0.5, (d) none of the above.

Q.110 When an electromagnetic wave is transmitted into a good conductor, it penetrates -

(a) unattenuated, (b) with fast attenuation,
(c) to zero depth, (d) none of the above.

Q.111 The Transverse Electric (TE) guided electromagnetic wave is characterized by -

(a) $E_{oz} = 0$, (b) $E_{oz} = 1$,
(c) $H_{oz} = 0$, (d) $H_{oz} = 1$.

Q.112 The Transverse Magnetic (TM) guided electromagnetic wave is characterized by -

(a) $E_{oz} = 0$, (b) $E0_z = 1$,
(c) $H_{oz} = 0$, (d) $H0_z = 1$.

Q.113 The Transverse Electric and Magnetic (TEM) guided electromagnetic wave is characterized by -

(a) $E_{oz} = 0, H_{oz} = 0$, (b) $E_{oz} = 1, H_{oz} = 1$,
(c) $H_{oz} = 0, E_{oz} = 1$, (d) $H_{oz} = 1, E_{oz} = 0$.

Q.114 Velocity of the following type guided waves is independent of frequency -

(a) TE, (b) TM,
(c) TEM, (d) none of the above.

Q.115 In a coaxial line for electromagnetic waves, the wave propagates through -

(a) concentric conductors, (b) space between conductors,
(c) both conductors and space, (d) none of the above.

Q.116 The value of wave impedance of a TEM wave propagating in a coaxial line, is -

(a) 112 ohms, (b) 377 ohms,
(c) 532 ohms, (d) none of the above.

Q.117 Usually a TEM can not be transmitted inside a hollow waveguide. It is true -

(a) except whwn wavelength << waveguide cross-section dimensions,
(b) always,
(c) rarely,
(d) none of the above.

Q.118 Wave guides can be used for transmission of TE and TM types of electromagnetic waves, only if free-space wavelength of the wave is -

(a) shorter than the width of the guide,
(b) shorter than twice the width of the guide,
(c) longer than the width of the guide,
(d) none of the above.

Q.119 Hollow wave guides for transmission of electromagnetic waves are inherently -

(a) low frequency device, (b) medium frequency device,
(c) high frequency device, (d) none of the above.

Q.120 The mode of propagation of a wave inside a hollow rectangular wave guide, is related to the wavelength and -

(a) width of the wave guide,
(b) length of the wave guide,
(c) attenuation in the wave guide,
(d) none of the above.

ABSTRACT

9.1. ELECTROMAGNETIC WAVES

The four basic equations of electromagnetic waves are the Maxwell equations :

$$\nabla \times E + \frac{\partial B}{\partial t} = 0$$

$$\nabla \times H + \frac{\partial D}{\partial t} = J$$

$$\nabla . D = \rho$$

and $$\nabla . B = 0$$

where E and H are electric and magnetic fields, such that :

$D = \varepsilon E$

$B = \mu H$

J and ρ are the current and charge densities.

Wave propagation

Taking curl of the first Maxwell equation :

$$\nabla \times (\nabla \times E) + \frac{\partial}{\partial t} (\nabla \times B) = 0$$

$$\nabla \times (\nabla \times E) = \nabla (\nabla . E) - \ {}^{2}E$$

For neutral material, $\rho = 0$, therefore $\nabla . E = 0$

Hence from second Maxwell equation :

$$\nabla^2 E = \varepsilon \mu_o \frac{\partial^2 E}{\partial t^2} + \sigma \mu_o \frac{\partial E}{\partial t}$$

This represents the propagation of wave with dissipation. Its general solution can be taken as :

$E = E_o \exp [i(k.r - \omega t)]$

so that $-k^2 = -\varepsilon \mu_o \omega^2 - \sigma \mu_o i \omega$

$$= \frac{\omega}{c} \left(\varepsilon_r + \frac{\sigma \mu_o i}{\omega} \right)^{1/2}$$

where $c = \sqrt{\frac{\varepsilon_o}{\mu_o}}$

and ε_r is the relative dielectric constant.

For free space $e_r = 1$ and $\sigma = o$

$k = \omega / c$

For a medium, phase velocity ω/k is modified by dividing c by a complex refractive index, given by :

$$n_r = n''_r + n''_r = \left(\varepsilon_r + \frac{\sigma\mu_o i}{\omega}\right)^{1/2}$$

$$\text{Hence } k = \frac{n'_r\omega}{c} + 1\, n''_r \frac{\omega}{c}$$

$$\text{Therefore } E = E_o^{\exp} \left[\iota\omega\left(\frac{n'_r t}{c} - t\right)\right] \exp\left(\frac{-n''_r\, wt}{c}\right)$$

Hence the velocity of wave is reduced from c/c'_r and its amplitude damped exponentially by a fraction exp $(-2\pi\, n''_r/n'_r)$ per wave length. This damping is associated with the absorption of electromagnetic energy. The absorption coefficient is described by the absorption of the intensity :

$$\alpha = \frac{2n''_r\omega}{c}$$

Note that in the absence of absorption $n'_r = n_r$ the refraction index will be simply n_r. For any material n_r can be measured and it provides information as n''_r.

On the similar lines, as the electric field, magnetic field may also be analysed :

$$\Delta^2 H = \varepsilon\mu_o \frac{\partial^2 H}{\partial t^2} + \sigma\mu_o \frac{\partial H}{\partial t}$$

which has propagating solution similar to electric field ·

$$H(r, t) = H_o \exp[i(k.r. - \omega t)]$$

Now energy density associated with electric and magnetic field is given by :

$$E_d = \tfrac{1}{2}(E.D. + H.B)$$

while the power density associated with radiation is given by the poynting vectors :

$$S = E \times H$$

To analyse the changes in propagation, consider a plane through which wave is passing. If z is normal to this plane dividing the medium into two regions with different material parameters, the normal components of D and B across this plane are related by :

$$[D(z^+) - D(z^-)].Z = \sigma_s$$

and $$[B(z^+) - B(z^-)]\,.\,Z = 0$$

where Z is a unit vector.

and the tangential components are related by :

$$Z \times [E(z^+) - E(Z^-)] = 0$$

$$Z \times [H(z^+) - H(Z^-)] = J_s$$

where J_s is the current density flowing along the surface.

If an electromagnetic plane wave is travelling along z-axis, the E and B fields are in the x-y plane and the electric and magnetic field components are given by :

$$E_y = \frac{\omega}{k} .B_x$$

$$E_x = -\frac{\omega}{k} B_y$$

Uniform plane wave

When electric field is independent of Y and Z and is a function of x and t only, the wave is said to be a uniform plane wave.

For such a wave $\nabla .E = \frac{1}{e} \nabla .D = 0$

$$H_x = \sqrt{\frac{\varepsilon}{\mu}} E_y$$

and $\frac{E}{H} = \sqrt{\frac{\mu}{\varepsilon}}$

Wave propagation in a dielectric

When $\sigma = 0$, intrinsic independence of dielectric $\pm\frac{\mu}{\varepsilon}$

Wave propagation in a dielectric

Velocity $v = \frac{w}{\beta} = \sqrt{\left(\frac{2\omega}{\mu\sigma}\right)}$

Intrinsic impedance $\eta = \sqrt{\frac{w\mu}{\sigma}} 45$

since α and β very large, velocity is very low and the phase shift per unit length is also large.

The depth of penetration $\delta = \frac{1}{\alpha} = \sqrt{\frac{2}{w\mu\sigma}}$

Wave polarization

Polarization of a uniform plane wave refers to the time-varying behaviour of the electric field strength vector at some fixed point in space.

Consider a wave travelling in z-direction with E and H vectors lying in x-y plane.

If $E_y = 0$ and only E_x is present, the wave is said to polarized in x-direction. If both E_x and E_y are present and angle $\tan^{-1} \dfrac{E_y}{E_n}$ is constant, the wave is linearly polarized. If E_x and E_y are not in phase, the resultant end-point will be an ellipse, the wave is said to be elliptically polarized. If $E_x = E_y$ and they have a 90° phase difference, the wave is said to be circularly polarized.

Reflection and refraction of waves

When a plane wave travelling in one medium impinges upon the boundary of a second medium having a different dielectric constant, permeability or conductivity, the wave is generally, partially transmitted and partially reflected. A perfect conductor totally reflects the wave.

When the wave obliquely reaches the surface of a dielectric making an angle θ_1 from the normal to the boundary, the refracted wave will make angle θ_2 with the normal to the boundary, such that :

that $$\frac{\text{Sin } \theta_1}{\text{Sin } \theta_2} = \sqrt{\frac{\varepsilon_2}{\varepsilon_1}} \quad \text{(Snell's law)}$$

and the reflected wave, will make angle of θ_3, such that

$$\theta_1 = \theta_3$$

When a vertically polarized wave reaches a boundary surface making an angle θ_1 with the normal, such that :

$$\tan \theta_1 = \sqrt{\frac{\varepsilon_2}{\varepsilon_1}}$$

there will be no reflection of the wave. This angle is known as Brewsters angle.

If medium 1 is denser than the medium 2 and the angle of incidence is such that :

$$\sin \theta_1 > \sqrt{\frac{\varepsilon_2}{\varepsilon_1}}$$

there will be a total reflection of the wave.

Surface impedance

Surface impedance of a conductor is given by :

$$Z_s = \frac{E_{tan}}{J_s}$$

where E_{tan} is the electric field strength parallel to, and the surface of the conductor and J_s is the linear current density, flowing due to E_{tan}.

$$Z_s = \frac{V}{\sigma}$$

For a thick conductor, surface impedance is equal to characteristic impedance. For good conductors :

$$Z_i \approx \frac{\omega\mu}{\sigma} < 45^o$$

surface resistance $R_s = \sqrt{\frac{w\mu}{2\sigma}}$ ≈ surface reactance

since depth of penetration is $\delta = \sqrt{\frac{2}{\omega\mu\sigma}}$

$$R_1 = \frac{1}{\sigma\delta}$$

Poynting theorem

The vector product P = E × H at any point is a measure of the rate of energy flow per unit area at that point. The direction of flow is perpendicular to E and H, in the direction of vector E × H.

Transverse electric wave (TE wave) and TM wave

If a wave has a component of H in the direction of propagation, but no component of E in this direction, the wave is send to be TE wave. On the other hand if in the direction of propagation component of only E exists, the wave is said to be a transverse magnetic (TM) wave.

Cut off frequency

For every guided wave, there exists a cut off frequency given by :

$$f_c = \frac{m}{2a\sqrt{(\mu\varepsilon)}}$$

below which there will be no wave propagation. Where m = 1, 2, 3, ... and a is distance between the guiding plates.

TEM (Transverse electromagnetic) waves

If the components of both E and H in the direction of propagation are zero, the wave is send to be a TEM wave. A TEM wave can not exist in a wave guide of any shape.

Waves in wave guides

(i) Characteristic impedance

Wave impedance Z_z (TM) $= \sqrt{\frac{\mu}{\varepsilon}}\,[1 - (w_c/w)^2]$

$$= \eta\,\sqrt{1 - (w_c - w)^2}$$

$$Z_z\,(TE) = \sqrt{\frac{\eta}{1 - (w_c/w)r^2]}}$$

For TEM wave in parallel plates
Z_z (TEM) = η

(ii) Attenuation factor

$$\alpha = \frac{\text{Power lost per unit length}}{\text{2x power transmitted}}$$

9.2. TELECOMMUNICATIONS

In electrical engineering telecommunication is referred to :

(i) Sorting, processing and storage of information,

(ii) Transmission and further processing for this purpose, of the information, and

(iii) Reception of the information and reprocessing, such as decoding, storage and interpretation.

Radiofrequency waves play an important role in transmission of the information as its carriers from one place to another. These waves are classified as :

(i) Very low frequency (VLF)	10-30 kHz
(ii) Low frequency (LF)	30-300 kHz
(iii) Medium frequency (MF)	300 kHz-3MHz
(iv) High frequency (HF)	3-30 MHz
(v) Very high frequency (VHF)	30-300 MHz
(vi) Ultra high frequency (UHF)	300-3000 MHz
(vii) Super high frequency (SHF)	3-30 GHz

The transmission of messages is by these waves and for this purpose these waves are modulated with the message waves. Usual carrier wave frequencies in use, are :

(a) Speech and other	10-15 kHz
(b) Video data	Upto 4.5 MHz

Modulation

An alternating voltage wave is completely defined as :

$v = A \cos (\omega t + \theta_o)$

The angular position of a rotating phaser is defined as :

$\theta = \int w dt + \theta_o$

where ω = angular velocity = $\frac{d\theta}{dt}$

Therefore $v = A \cos [\omega dt + \theta_o]$

This is a preset or anticipatative wave, which has no information content. This wave can be made to carry an information in its characteristic values which are its amplitude and the phase angle. Hence the process of modulation of the basic carrier wave, can be put into two classes :

(i) *Amplitude modulation* (AM) in which the carrier wave amplitude is made to vary in time in accordance with the information wave.

(ii) *Angle modulation,* in which the instantaneous value of the carrier wave angle is changed. This process may be subdivided into *Frequency Modulation* (FM), wherein ω varies in time in accordance with the modulating signal amplitude, and the *Phase modulation* (PM), in which angle θ is varied by the modulating signal.

Another type of modulation system involves turning the carrier ON and OFF and hereby generating pulses of radiofrequency. This method samples the informational signal at a high rate and the characteristics of the pulses is varied in accordance with the amplitudes of the signal samples. These systems include *Pulse-amplitude modulation* (PAM), *Pulse-duration modulation* (PDM), and *Pulse-code modulation* (PCM).

Amplitude modulation

In amplitude modulation, the angular velocity ω of the carrier wave is kept constant as carrier frequency $f_c = \omega_c/2\pi$. This fixes the carrier wave as :

$v_c = A \cos (w_c t + \theta_o)$

If the signal, carrying the information is :

$v_m = v_m \cos \omega_m t$

The amplitude of v_c is made :

$$A = v_c + v_m \cos \omega_n t$$

$$= v_c \left(1 + \frac{v_m}{v_c} \cos \omega_m t)\right)$$

$$v = v_c (1 + m_a \cos w_m t) \cos w_c t$$

where the constant phase angle θ_o is droppled and m_a, known as the modulation factor is defined as :

$$m_a = \frac{V_m}{V_c}$$

which is not allowed to exceed unity or 100% per cent.

Frequency modulation

In frequency modulation, the amplitude of the carrier wave is kept constant but its angular velocity ω is made to vary in accordance with the amplitude of the informational signal. The modified velocity may be written as :

$$\omega = \omega_o + k_f V_m \cos \omega_m t$$

where ω_o = $2\pi f_o$ is the center angular velocity corresponding to the center frequency f_o.

ω_m = modulation angular frequency

$k_f V_m$ = degree of frequency variation proportional to the modulating amplitude.

Thus, the modulated wave is :

$$v = V_o \cos\left(\omega_o t + \frac{k_f V_m}{\omega_m} \sin \omega_m t\right)$$

where $\frac{k_f w}{\omega m}$ is the frequency deviation from the centre frequency and is designated as Δf. The ratio of maximum deviation to the maximum modulating frequency is called the deviation ratio of the system, given by:

$$m_f = \frac{\Delta f_{max}}{f_{m\ max}}$$

making the final modulated wave as :

$$v = V_o \cos(\omega_o t + m_f \sin \omega_m t)$$

Thus the frequency deviates from the center frequency in proportion to the amplitude of the modulating signal.

Phase modulation

In phase modulation both the amplitude and the angular velocity are kept constant and the phase angle, which was constant at θ_o, is made to vary in accordance with the modulating signal amplitude. The modulated wave takes the form :

$$v = V_c \cos(\omega_o t + \phi)$$

where $\phi = \phi_o + k_p V_m \cos \omega_m t$

where k_p is the phase angle variation per unit signal amplitude. Thus :

$$v = V_c \cos(\omega_o t + k_p V_m \cos \omega_m t)$$
$$= V_c \cos(\omega_{o7} t + \phi_d \cos w_m t)$$

where $\phi_d = k_p V_m$, known as phase deviation.

Definitions

Parsevals power theorem

The average power of a periodic signal is equal to the sum of the squared amplitudes of all the harmonic components of the signal.

Conditions of distortionless transmission

(i) The amplitude response is constant for all frequencies,

(ii) The phase shift is linear with frequency, passing through a zero or a multiple of π at zero frequency.

Amplitude distortion

When the amplitude response of a system is not constant for all frequencies, different frequency-components of the input signal are transmitted with different gain or attenuation. The effect is the amplitude distortion.

Phase distortion

When the phase response of the channel is not linear with frequency, then components of input signal in different narrow bands of frequencies are phase shifted at the output, resulting in the phase distortion.

Aliasing effect

It is related to sampling technique. It is the phenomenon of a high-frequency component in the spectrum of the original signal seemingly taking on the identity of a lower frequency in the spectrum of its sampled version.

Pulse-amplitude modulation

The amplitudes of regularly speed rectangular pulses vary with the instantaneous sample values of a continuous message signal in a one-to-one fashion.

Pulse-duration modulation (PDM)

The samples of the message signal are used to vary the duration of the individual pulses of the carrier wave. This modulation is also known as pulse-width modulation or pulse-length modulation.

Pulse-position modulation (PPM)

The position of a pulse relative to its unmodulated time of occurrence is varied in accordance with the message signal.

Pulse-code modulation (PCM)

The message signal is sampled and the amplitude of each sample is rounded off to the nearest one of a finite set of allowable values, so that both time and amplitude are in discrete form. This allows the message to be transmitted by means of coded electrical signals.

Channel capacity

The maximum rate of reliable information transmitted over a channel. It is governed by the capacity theorem, which states :

If $rH < C$, then there exists a coding sentence such that the output of the source can be transmitted over the channel with an arbitrarily small probability of error.

10

Power Electronics

Q.1 With the rise in temperature, the resistance of a semiconductor-

(a) rises, (b) falls,

(c) remains steady, (d) is unpredictable.

Q.2 A diode can be used for conducting current in the direction from cathode to anode, without any damage to it.

(a) true, (b) false,.

Q.3 The potential barrier of an ordinary diode is of the order -

(a) 0.06 volts, (b) 0.6 volts,.

(c) 2.0 volts, (d) none of the above.

Q.4 In a silicon-controlled rectifier, the gate is always made of the -

(a) N layer, (b) any P layer,

(c) middle P layer, (d) none of the above.

Q.5 The driving pulse to a silicon-controlled rectifier is applied between -

(a) gate and anode, (b) gate and cathode,.

(c) anode and cathode, (d) none of the above.

Q.6 Once a silicon-controlled rectifier is switched on, to remain in conductive mode, it needs -

(a) gate pulse only,

(b) forward bias voltage only,

(c) gate pulse and forward bias voltage both,

(d) any of the gate pulse and the forward bias voltage.

Q.7 T switch off a silicon-controlled rectifier, from the conduction mode -

(a) only gate pulse has to be withdrawn,

(b) only forward bias has to be withdrawn,

(c) both gate pulse and forward bias need be withdrawn,.
(d) negative gate pulse has to be applied.

Q.8 Minimum level of gate voltage and current required to turn on the thyristor is a function of junction temperature, higher the temperature -

(a) lower the required puse, (b) bigger the required pulse,

Q.9 Gate characteristic of a silicon-controlled rectifier shows higher the applied voltage -

(a) lower will be the current required,.
(b) higher will be the current required,
(c) current reqirement remains steady.

Q.10 A triac is equivalent to two thyristors connected together in -

(a) series, (b) paralel,
(c) inverse-series, (d) inverse-parallel.

Q.11 A triac is used to control of -

(a) ac current flow, (b) dc current flow,
(c) ac voltage blocking, (d) dc voltage blocking.

Q.12 A gate turn-off (GTO) thyristor, can be switched off by -

(a) withdrawal of the gate current only,.
(b) withdrawal of the anode-cathode bias only,
(c) withdrawal of both gate current and the anode-cathode bias,
(d) none of the above.

Q.13 In a power transistor the unidirectional current between the emitter and the colector can be controlled by -

(a) anode-cathode voltage, (b) base-emitter voltage,
(c) base-emitter current, (d) none of the above.

Q.14 For the some applied voltage and the load, in a thyristor-controlled circuit, the curent at any moment is predecided, while in a power transistor-controoled circuit, the current can be controlled by -

(a) emitter control, (b) collector control,
(c) base control, (d) none of the above.

Q.15 A power MOSFET is a voltage controlled device which means that its drain-source current is controlled by -

(a) gate-source current, (b) gate-source voltage,.
(c) gate-drain current, (d) none of the above.

Q.16 A fully-controlled thyristor bridge circuit can be used for power flow in either direction, in forward direction the circuit is called a convertor, while for reverse power flow the circuit called -

(a) a rectifier, (b) an invertor,
(c) a revertor, (d) none of the above.

Q.17 By a bridge circuit with one thyristor and one diode in each leg, the output dc voltage is -

(a) controllable in upper-half range,
(b) controllable in lower-half range,
(c) controllable in full range,.
(d) none of the above.

Q.18 All half-controlled and uncontrolled rectifier circuits have a reverese connected diode across the load., known as free-wheeling or commutating diode. It bypasses the load voltage when it -

(a) goes negative momentarily, (b) crosses the normal rating,
(c) fast reversed, (d) none of the above.

Q.19 In a half-wave single phase rectifier circuit, for ac peak voltage of V_m volts, the output voltage is -

(a) 1.44 V_m, (b) V_m / π,.
(c) $V_m / \sqrt{2}$, (d) none of the above.

Q.20 For a bi-phase half-wave rectifier circuit, if V_m is the peak voltage of each phase, the output voltage will be -

(a) $2 V_m/\pi$, (b) $2 V_m/\sqrt{2}$,
(c) $\sqrt{2}. V_m$, (d) none of the above.

Q.21 For a single-phase full-wave rectifier circuit, if V_m is the peak input voltage, the output voltage will be -

(a) $2 V_m/\pi$, (b) $2 V_m/\sqrt{2}$,
(c) $\sqrt{2}. V_m$, (d) none of the above.

Q.22 For a three phase half-wave rectifier circuit, if input phase peak is V_m volts, the output dc mean voltage will be -

(a) $3 \sqrt{3} V_m / 2 \pi$,. (b) $3 V_m / \pi$,
(c) $3V_m / \sqrt{2}$, (d) none of the above.

Q.23 For a six-phase half-wave rectifier circuit, if input phase peak is V_m volts, the output dc mean voltage will be -

(a) $3 V_m / 2 \pi$, (b) $3 V_m / \pi$,
(c) $3 V_m / \sqrt{2}$, (d) nonc of tho above.

Q.24 For a 3-phase full-wave rectifier bridge, if the input line peak is V_m volts, the output mean dc voltage will be -

(a) $3 V_m/\pi$, (b) $3 V_m /\sqrt{2}$,
(c) $\sqrt{3} V_m /2$, (d) none of the above.

Q.25 When for a circuit the input is a dc voltage and output a dc controlled voltage, the circuit is called -

(a) dc/dc controller, (b) dc chopper,.
(c) dc convertor, (d) none of the above.

Q.26 In a dc chopper circuit, the input voltage is connected to the load with ON/OFF electronic switches, the mean output voltage depends upon -

(a) ON duration / OFF duration,
(b) ON duration / cycle duration,
(c) OFF duration / cycle duration,
(d) none of the above.

Q.27 The process of producing an output perodic wave at a frequency different from the frequency of the input periodic wave is known as -

(a) inversion, (b) cycloconversion,.
(c) conversion, (d) none of the above.

Q.28 In cycloconvertor circuits, two convertor circuits provide the output and these convertors are operated in -

(a) cascade, (b) inverse-cascade,
(c) parallel, (d) inverse-parallel.

Q.29 In cycloconvertors, as compared to the input frequency, the output frequency -

(a) is always less, (b) can be more or equal,
(c) is always more.

Q.30 In envelop cycloconvertors, the ratio of input and output frequencies -

(a) can be continuously varied, (b) has to be an integer,.
(c) can be more than unity, (d) none of the above.

Q.31 A circuit converting dc input to a periodic wave output, is said to be -

(a) convertor, (b) invertor,.
(c) cycloconvertor, (d) none of the above.

Q.32 A dc motor could be controlled by a variable transformer and a diode rectifier circuit. The notable features of this scheme are (i) longer response time, (ii) good power factor, (iii) more harmonics.

(a) all the above are true, (b) only first two are true,.
(c) only last two are true, (d) none of the above.

Q.33 The advantages of a dc motor control through a controlled rectifier are (i) fast response, (ii) flexibility, (iii) high power factor, (iv) low harmonics.

(a) all the above are true, (b) only first two are true,.

(c) only first three are true, (d) none of the above is true.

Q.34 DC motor control with a convertor circuit leads to derating of the motor capacity, due to -

(a) inductance of the motor,

(b) instabilty of the speed response,

(c) harmonic contents of the supply,.

(d) none of the above.

Q.35 For dynamic braking of a convertor controlled dc motor, the convertor circuit -

(a) should be used as invertor, (b) has no role to play,.

(c) should be shortcircuited, (d) none of the above.

Q.36 For regenerative braking of a convertor controlled dc motor, the convertor circuit -

(a) should be used as invertor, (b) has no role to play,

(c) should be shortcircuited, (d) none of the above.

Q.37. DC motor fast-reversing armature control can be achieved by -

(a) half-controlled convertor, (b) fully-controlled convertor,.

(c) diode rectifiers, (d) none of the above.

Q.38 DC motor fast-reversing armature control can be achieved by a fully-controlled convertor and a set of contactors or two fully-controlled convertors connected in -

(a) cascade, (b) inverse-cascade,

(c) parallel, (d) inverse-parallel.

Q.39 The dc motor reversing-control can be achieved by field reversal also but this process is slow, as comparedto the armature reversal, because -

(a) high inductance of the armature,

(b) high inductance of the field winding,

(c) low power requirement of the field circuit,

(d) none of the above.

Q.40 Power factor of the current drawn from the ac source, in case of a convertor-controlled dc motor, is -

(a) constant,

(b) proportional to motor speed,.

(c) inversely proportional to motor speed,

(d) none of the above.

Q.41 For stablized speed control of DC motors through convertors, feedback of the speed can be introduced through the use of -

(a) RC network, (b) tachogenerator,.

(c) potential transformer, (d) none of the above.

Q.42 To limit current level of a convertor-controlled dc motor, in response to a speed change signal, current feedback is used through the use of -

(a) dc current transformers, (b) RC network,
(c) direct loopback, (d) none of the above.

Q.43 Speed control of a dc series motor is done through the use of -

(a) chopper circuit, (b) cycloconvertor circuit,
(c) convertor circuit, (d) none of the above.

Q.44 For induction motors, thyristorized ac voltage control can be beneficially utilized in -

(a) speed control, (b) soft starting,
(c) quick starting, (d) none of the above.

Q.45 For induction motor control through frequency control, it is desirablee that the output voltage of the invertor is kept -

(a) constant,
(b) proportional to frequency,.
(c) inversely proportional to frequency,
(d) none of the above.

Q.46 In induction motor speed control drive, the variable-frequency invertor may be fed from (i) a fully-controlled convertor, or (ii) a diode-rectifier and chopper, to keep output voltage proportional to frequency.

(a) all the above are true, (b) only (i) is true,
(c) only (ii) is true, (d) none of the above is true.

Q.47 The following type of invertor circuit for induction motor speed control is suitable -

(a) quasi-square wave invertor,
(b) pulse-width modulation invertor,
(c) constant voltage invertor,
(d) none of the above.

Q.48 For induction motor speed control, fixed voltage/frequency control is obtained by -

(a) voltage control with frequency following,.
(b) frequency control with voltage following,
(c) both voltage and frequency control,
(d) none of the above.

Q.49 For synchronous motor speed control, fixed voltage/frequency control is obtained by -

(a) voltage control with frequency following,
(b) frequency control with voltage following,.

(c) both voltage and frequency control,
(d) none of the above.

Q.50 In fixed voltage/frequency control for induction motor speed control, the motor torque is controlled through control of -
(a) invertor input voltage, (b) invertor frequency,
(c) invertor input current, (d) none of the above.

Q.51 In fixed voltage/frequency control for induction motor speed control, the speed regulation can be improved by -
(a) output voltage feedback, (b) dc link current feedback,
(c) tachometer feedback, (d) none of the above.

Q.52 In fixed voltage/frequency control for induction motor speed control, the curent limit can be achieved by -
(a) output voltage feedback, (b) dc link current feedback,.
(c) tachometer feedback, (d) none of the above.

Q.53 In fixed voltage/frequency control for synchronous motor speed control, the stability of the motor can be improved by -
(a) introducing a ramp function in the input signal,.
(b) dc link current feedback,
(c) tachometer feedback,
(d) none of the above.

Q.54 Reluctance and permanent magnet synchronous motors are usually controlled by fixed voltage/frequency control through -
(a) open loop,.
(b) closed loop through speed feedback,
(c) closed loop through voltage feedback,
(d) none of the above.

Q.55 THe four-quadrant motor control means controls of -
(a) motor speed, torque, voltage and current,
(b) motor speed and torque in both directions,.
(c) motor voltage and current in both directions,
(d) none of the above.

Q.56 For ac motor speed controls through constant voltage/frequency ratio, it is desirable that the invertor is fed at -
(a) constant dc link current, (b) constant line voltage,
(c) constant line current, (d) none of the above.

Q.57 In ac motor speed controls through constant voltage/frequency ratio, the regenerative braking may be adopted through -
(a) reversing the dc link current,
(b) reversing the dc link voltage,.

(c) rerversing the line voltage,
(d) none of the above.

Q.58 In ac motor control through cycloconvertors, with keeping the voltage/frequency ratio fixed, the main limitation is low output frequency of the cycloconvertor, which is about -
(a) equal to the input frequency,
(b) half the input frequency,
(c) one third of the input frequency,
(d) none of the above.

Q.59 In ac motor speed control with fixed voltage/frequency ratio through the use of transistorized diode/chopper/invertor circuits, the chopping rate can be much higher as compared to thyristor circuit. This results in saving on -
(a) size of main control components,
(b) line components,
(c) L-C filter components,
(d) none of the above.

Q.60 Slipring induction motors speed control through electronic methods involve -
(a) input voltage control, (b) rotor voltage injection,
(c) slip-energy recovery, (d) none of the above.

Q.61 Slip-energy recovery method of speed control of slipring induction motors is traditionally known as -
(a) Scherbius system, (b). Kramer system,
(c) Leblanc system, (d) none of the above.

Q.62 Slipring induction motor speed control, by electronic methods, requires conversion of rotor frequency to mains frequency and connected back to mains. This can be achieved by (i) cycloconvertor, (ii) rectifier-invertor, (iii) two full-convertors, (iv) two half-convertoors.
(a) all the above are true, (b) only first two are true,
(c) only (ii) and (iii) are true, (d) only last two are true.

Q.63 In slipring motor speed control method, super-synchronous speed can be achieved by -
(a) injecting power in rotor,
(b) blocking the flow of rotor-energy,
(c) incresing the input voltage,
(d) none of the above.

Q.64 By connecting a cycloconvertor in between the rotor and the mains of a running slipring induction motor -
(a) sub-synchronous speed control can be achieved,
(b) super-synchronous speed control can be achieved,.

(c) none of the above is true.

Q.65 In stepper motors, the reluctance of motor's magnetic path is different for different -
(a) poles, (b) phases,
(c) angular positions of the rotor, (d) none of the above.

Q.66 The stepper motors are used for precise -
(a) speed controls, (b) torque controls,
(c) position controls, (d) none of the above.

Q.67 Stepper motor rotors are -
(a) cylinderical with a winding,
(b) with salient pole winding,
(c) with salient poles without winding,
(d) none of the above.

Q.68 Stepper motors always have multi-phase winding, and its energization is done -
(a) alternate phases at a time, (b) one phase at a time,.
(c) all phases at a time, (d) none of the above.

Q.69 In a stepper motor, number of magnetic poles on the stator and the rotor -
(a) are always equal, (b) are always unequal,.
(c) may be equal, (d) none of the above.

Q.70 In hybrid type of stepper motors, two rotor disks are sandwitched with a permanent magnet. The torque being produced by -
(a) stator only, (b) rotor only,
(c) stator and rotor both, (d) none of the above.

Q.71 The frequency at which induction heating for heat treatment of metals can be done is around -
(a) 400 - 1000 kHz,. (b) 1 MHz - 10 MHz,
(c) 10 MHz - 20 MHz, (d) none of the above.

Q.72 Induction heating process can be used to heatup -
(a) insulators, (b) metals,
(c) nonmagnetic material, (d) none of the above.

Q.73 Induction heating works on the principle of -
(a) inductor, (b) capacitor,
(c) transformer, (d) none of the above.

Q.74 The electronic component used for high-frequency induction heating is usually -
(a) Vacuum tube, (b) power transistor,
(c) thyristor, (d) none of the above.

Q.75 Thyristorized invertors can be used for medium frequency induction heating for the purpose of -
(a) metal melting,.
(b) plywood pasting,
(c) surface heat treatment of metals,
(d) none of the above.

Q.76 In induction heating, the heat is mostly concentrated on the surface of the metal being heated due to -
(a) Hall's effect, (b) skin effect,.
(c) Lenz's law, (d) none of the above.

Q.77 The heating of metals by the process of induction heating is caused by -
(a) magnetic effect on metals, (b) electric arc,
(c) electric current in the metal, (d) none of the above.

Q.78 For an induction heating system, if the heated metal is nonmagnetic, its heating, for all conditions remaining the same, as compared to that of a magnetic metal, will be -
(a) faster, (b) slower,
(c) same,

Q.79 As compared to nonmagnetic metals, magnetic metals are heated faster by induction heating due to -
(a) higher conductivity of magnetic materials,
(b) higher resistance of magnetic materials,
(c) higher magnetic induction in magnetic materials,
(d) none of the above.

Q.80 A cycloconverter can not be used for providing induction heating power because -
(a) it is too costly,
(b) it provides frequencies lower than that required,
(c) its voltage is too low for induction heating,
(d) none of the above.

Q.81 Dielectric heating works on the principle of -
(a) inductor, (b) capacitor,
(c) transformer, (d) none of the above.

Q.82 Dielectric heating can be used for -
(a) heating of plywoods, (b) steel plates,
(c) water, (d) none of the above.

Q.83 In dielectric heating, higher the frequency, for the same heating-
(a) higher will be the applied voltage,
(b) lower will be the applied voltage,.

(c) same will be the voltage,

Q.84 Dielectric heating circuit works at -
(a) leading power factor, (b) lagging power factor,
(c) unity power factor,

Q.85 The usual induction heating circuits work at -
(a) leading power factor, (b) lagging power factor,
(c) unity power factor,

Q.86 Lasers cab used for heating -
(a) only metals, (b) only nonmetals,
(c) metals and non-metals both, (d) none of the above.

Q.87 The proprty of coherence of lassers, as compared to that of light, is -
(a) higher, (b) lower,
(c) same,

Q.88 In arc furnaces, carbon/graphite electrodes are used due to -
(a) their high resistance,
(b) their high melting point,
(c) their high thermal conductivity,
(d) none of the above.

Q.89 In nonlinear resistors, -
(a) the current drops with voltage rise,
(b) the resistance decreases as voltage is increased,.
(c) the conductivity drops as voltage increses,
(d) none of the above.

Q.90 For high voltage discharge testing, high voltage is obtained by -
(a) transformer-rectifier circuit,
(b) transformer-thyristor circuits,
(c) voltage doubler using rectifier-capacitor circuits,.
(d) none of the above.

Q.91 The spark-erosion process is used -
(a) to digout shapes in the conducting metals,.
(b) to shape insulating solids,
(c) ionize gases,
(d) none of the above.

Q.92 In spark-erosion process mechanical vibrations and electrical pulses are synchronized to create miniature sparks between the electrodes at about -
(a) 1000 volts, (b) 500 volts,
(c) 100 volts, (d) 2 volts,.

Q.93 Laser welding is possible at spots, inaccessible to normal tools and hands, but accessible to -

(a) light ray, (b) flow of electricity,
(c) the heat, (d) none of the above.

Q.94 What is the wave length range of optic rays -

(a) 0.01 μm to 100 μm, (b) 0.01 mm to 100 mm,
(c) 0.01 m to 100 m, (d) none of the above.

Q.95 What is the frequency range of optic waves -

(a) 3×10^{10} - 3×10^{12} Hz, (b) 3×10^{13} - 3×10^{15} Hz,.
(c) 3×10^{16} - 3×10^{18} Hz, (d) none of the above.

Q.96 Nondoped semiconductors such as Galium arsenide, Cadmium sulphide and cadmium selenide can be used to emit lasers at -

(a) high temperatures, (b) low temperatures,
(c) any temperature.

Q.97 The injection type lasers are generated by p-n junction of semiconductors such as -

(a) Ga-As chip, (b) any p-n junction,
(c) none of the above.

Q.98 Plasma of a ionic gas means -

(a) highly concentrated gas with overheated electrons,
(b) highly rarefied gas with overheated electrons,.
(c) highly rarefied gas with supercooling,
(d) none of the above.

Q.99 Chemical lasers are generated by exothermal chemical reactions such as that of (i) deuterium and fluoride, (ii) hydrogen and fluoride.

(a) all the above are true, (b) only (i) is true,
(c) only (ii) is true, (d) none of the above.

Q.100 A mixture of 8% carbondioxide, 90% nitrogen and 2% water moisture, when heated to 1500 K and dynamically fast cooled by diffusion through a nozzle, produces Carbondioxide lasers.

(a) true, (b) false.

Q.101 The ruby lasers are produced by ruby crystal, which is -

(a) aluminium trioxide, (b) zinc sulphide,
(c) lead phosphate, (d) none of the above.

ABSTRACT

10.1. POWER ELECTRONIC DEVICES

(i) Thyristor

One of the most important solid-state power devices is thyristor, or silicon-controlled-rectifier, as it was known before gettint the name thyristor. The thyristor is a semiconductor switch either of the pnpn or npnp type, whose bi-stable action depends on regenerative internal feedback. The four-layer device is usually of silicon though germanium has also been used. Its analytical construction and working is shown in Fig. 10.1. The structure is visualized as consisting of two transistors, a pnp and npn interconnected to form a regenerative feedback pair. For indidivual current gains h_{fe1} and h_{fe2} of the two transistor, the current gain for the internal feedback loop is $h_{fe1} \times h_{fe2}$. If I_{co1} and I_{co2} are collector to base leakage currents of the npn and pnp sections respectively, then :

for the pnp section : $I_{c1} = h_{fe1} (I_{c2} + I_{co1}) + I_{co1}$

for the npn section : $I_{c2} = h_{fe2} (I_{c1} + I_{co2}) + I_{co2}$

Total anode - to-cathode current $I_a = (I_{c1} + I_{c2})$

$$\text{therefore } I_a = \frac{(1 + R_{fe1})(1 + hf_{fe2})(I_{co1} + I_{co2})}{1 - (h_{fe1})(h_{fe2})}$$

With bias applied between anode and cathode with anode positive the device is said to be in forward blocking or high-impedance off state. The switch to the low-impedance 'on' state is initiated by raising the loop gain to unity. By this the device starts, regeneration, each transistor driving the other to saturation. Once in saturation, all junctions assume a forward bias, and the total potential drop across the device approximates to that of the single junction. Anode current is then limited by external circuit only.

To turn-off the hyristor in a minimum time, it is necessary to apply a reverse voltage and under this condition, the charge carriers in the vicinity of two end junctions will diffuse in these junctions and result in a reverse current in the external circuit. The voltage across thyristor will remain 0.7 V positive till an appreciable reverse current flows which removes the holes and electrons in the vicinity of two end junctions. At this the reverse current will cease and the junction will assume a blocking state. The turn-off time is usually of the order of 10-15 ms.

The main difference between a transistor and a thristor is that in transistor the conduction can be stopped at any point in the cycle because of no internal regeneration and current gain less than unity. In thyristor, conduction can stop only when the current is zero.

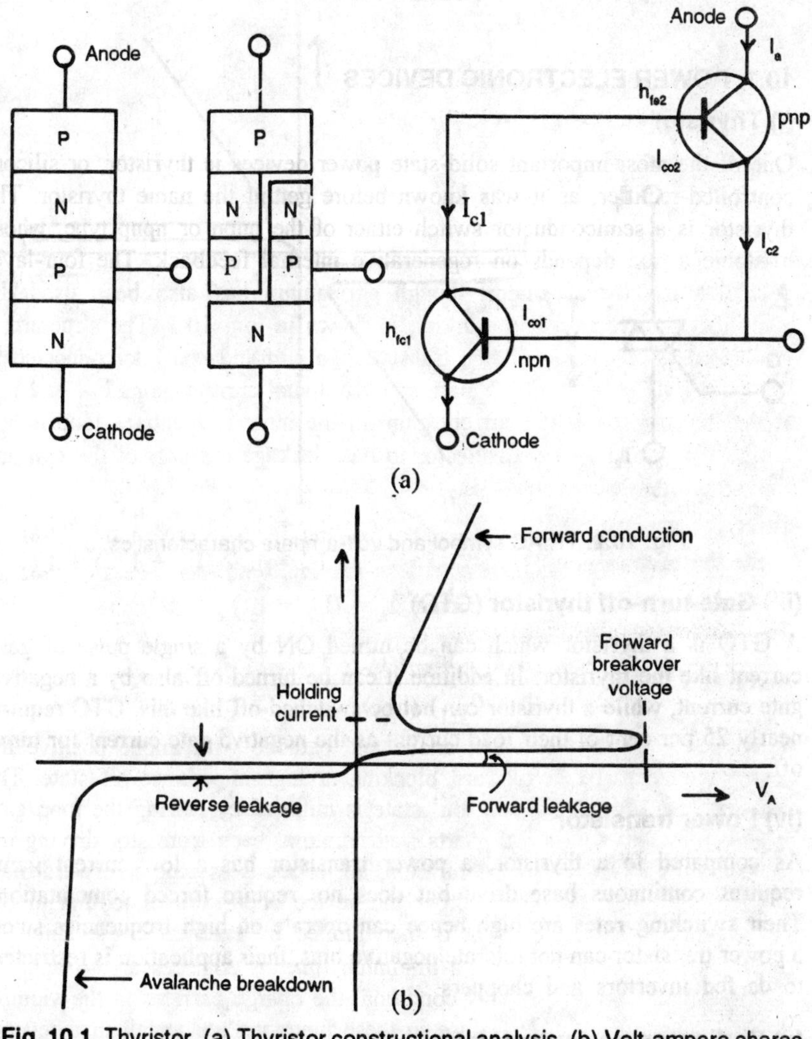

Fig. 10.1. Thyristor. (a) Thyristor constructional analysis. (b) Volt-ampere characteristics.

(ii) TRIAC

Functionally, a triac is an anti-parallel adjoint of two thyristors. Hence it can conduct in both directions. The volt-ampere characteristics of a triac is shown in Fig. 10.2. For flow of current in reverse direction, the applied voltage and the gate pulse has to be reversed.

A triac is economical than two thyristors but is slow in turning off. It is used for phase control with a resistive load for supply frequencies of 60 to 400 Hz. Triac voltage and current ratings are also smaller than those of thyristors.

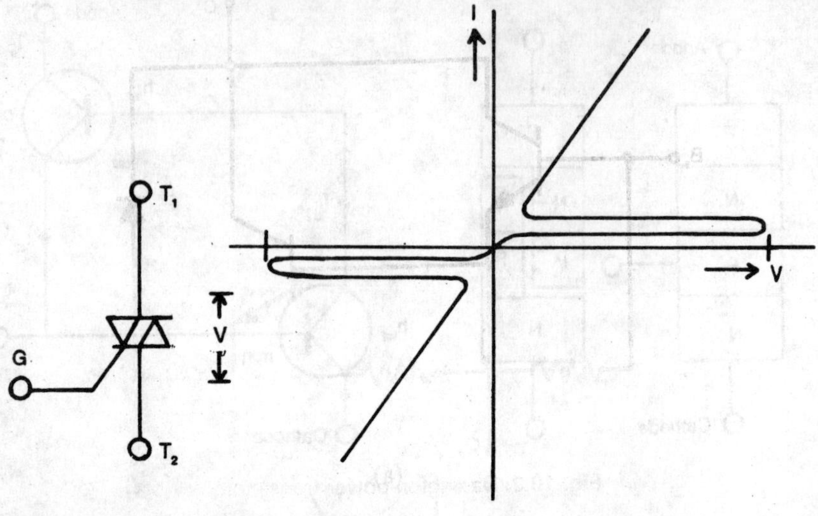

Fig. 10.2. TRIAC symbol and volt-ampere characteristics.

(iii) Gate-turn-off thyristor (GTO)

A GTO is a thyristor which can be turned ON by a single pulse of gate current like the thyristor. In addition it can be turned off also by a negative gate current, while a thyristor can not be switched-off like this. GTO require nearly 25 per cent of their load current as the negative gate current for turn-off.

(iv) Power transistor

As compared to a thyristor, a power transistor has a low current gain, requires continuous base drive but does not require forced commutation. Their switching rates are high hence can operate on high frequencies since a power transistor can not tolerate negative bias, their application is restricted to dc fed invertors and choppers.

(v) Darlington power transistor

This is a combination of two power transistors in cascade connection for a large common-emitter current gain. Their switching frequency is lower and conduction drop higher than the ordinary power transistors. Fig. 10.3 shows the internal connections of a Darlington transistor.

(vi) Power MOSFET

For low power high-frequency applications it has outpaced the power transistor in many respects. Presently, it is constructed in planar (D-MOS) design, shown in Fig. 10.4(a). Fig. 10.4(b) shows its circuit symbol. Its

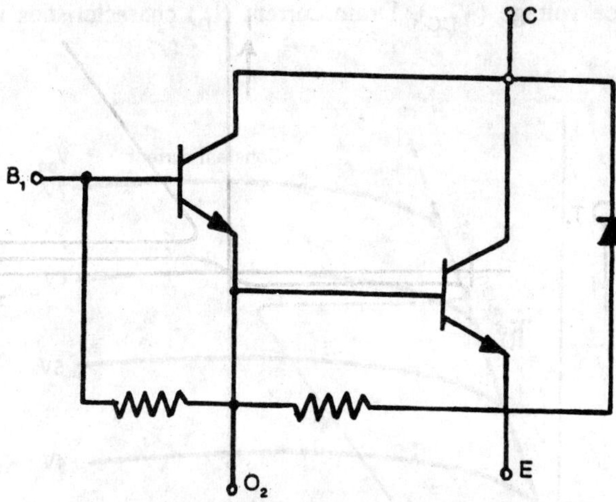

Fig. 10.3. Darlington power transistor.

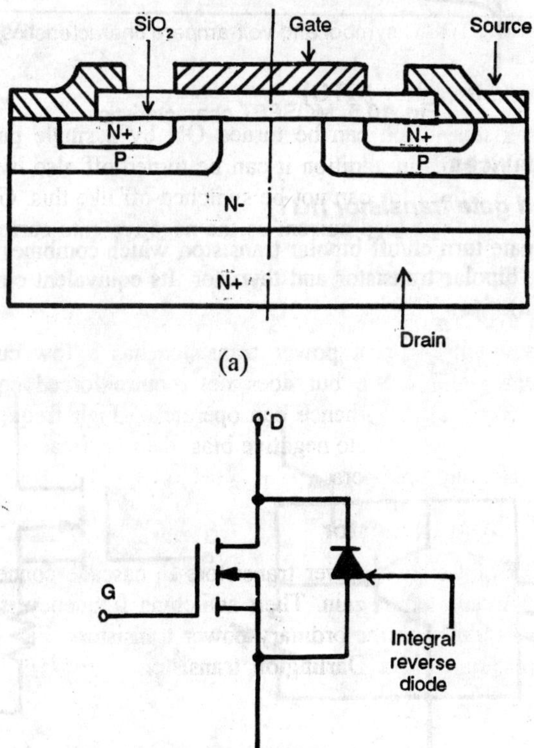

Fig. 10.4. Power MOSFET (N-type). (a) Planar construction (D-MOS). (b) Circuit-symbol.

drain source voltage (V_{DC}), Drain current (I_D) characteristics is shown in Fig. 10.5.

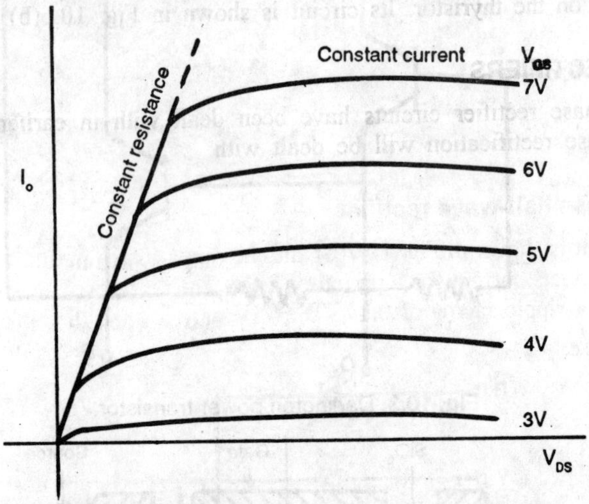

Fig. 10.5. MOSFET charactaristics.

(v) Hybrid devices

(a) Insulated gate transistor (IGT)

It is a MOS gate turn on/off bipolar transistor, which combines the attributes of MOSFET, bipolar transistor and thyristor. Its equivalent circuit is shown in Fig. 10.6(a).

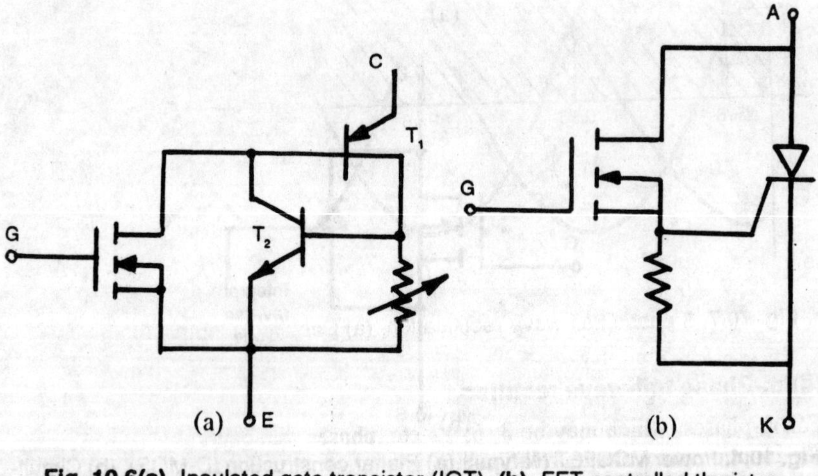

Fig. 10.6(a). Insulated gete transistor (IGT). **(b).** FET-controlled thyristor.

(b) FET-controlled thyristor

It is parallel combination of a MOSFET and a thyristor. It has the input impedence of FET and conduction drop of thyristor. The MOSFET current switches on the thyristor. Its circuit is shown in Fig. 10.6(b).

10.2. RECTIFIERS

Single-phase rectifier circuits have been dealt with in earlier. Here only three-phase rectification will be dealt with.

(i) 3-phase half-wave rectifier

The circuit is shown in Fig. 10.7(a) and the output wave in Fig. 10.7(b). Each diode conducts current for a period of $2\pi/3$. The output voltage will be 0.827 V_m with a ripple factor of 0.17. The PIV across each diode will be $\sqrt{3}V_m$ or 2.09 V_{dc}.

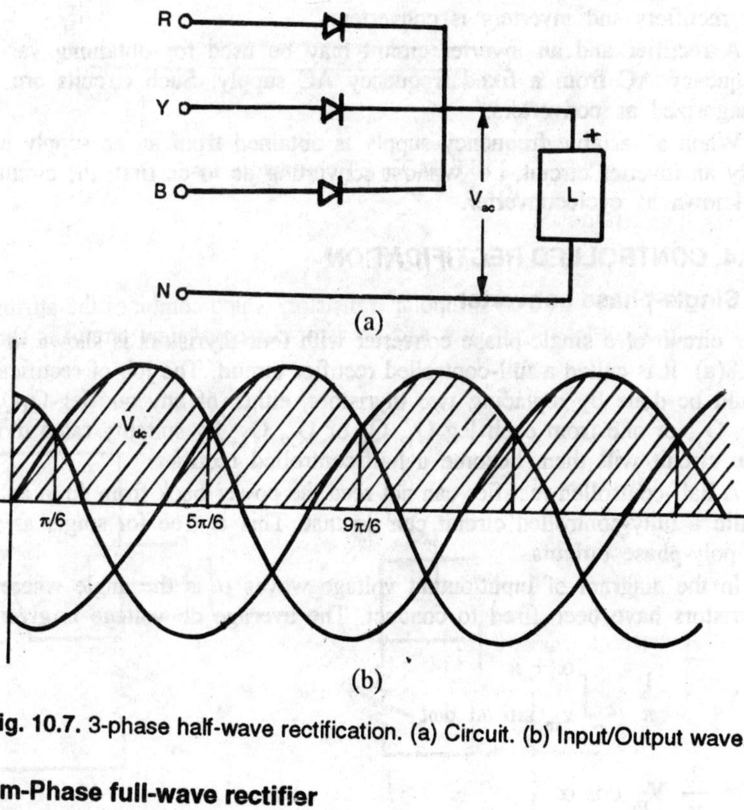

Fig. 10.7. 3-phase half-wave rectification. (a) Circuit. (b) Input/Output wave.

(ii) m-Phase full-wave rectifier

For m-phases, which may be 3, 6, 12 etc. phases, the generalized expression for input/output voltage relationship is :

$$\frac{V_{dc}}{V_m} = \frac{m}{p} \sin\frac{\pi}{m}$$

and power delivered to a resistive load is given by :

$$P_{dc} = \frac{V_m^2}{R}\left(\frac{m}{\pi} \sin \frac{\pi}{m}\right)^2$$

Table 10.1 gives various rectification parameters for various circuit configurations.

10.3. CONVERTERS AND CYCLOCONVERTERS

Phase time-controlled switching of thyristors is used for getting variable dc voltage from ac supply. Such circuits are known as controlled rectifier. The same circuits with some modifications could be used for conversion from dc to ac supply. Such circuits are known as invertors. The general term used for rectifiers and inverters is converters.

A rectifier and an inverter circuit may be used for obtaining variable frequency AC from a fixed frequency AC supply. Such circuits are also catagorized as converters.

When a variable frequency supply is obtained from an ac supply using only an inverter circuit, i.e. without converting ac to dc first, the circuitary is known as cycloconverter.

10.4. CONTROLLED RECTIFICATION

(i) Single-phase converter

The circuit of a single-phase converter with four thyristors is shown in Fig. 10.8(a). It is called a full-controlled rectifier circuit. The job of rectification could be done by replacing two thyristors, either of any one leg-Q_1Q_4 or Q_2, Q_3; or one from each leg Q_1, Q_3 or Q_2, Q_4; by suitably rated diodes. The circuit will then, become a half-controlled rectifier.

A half-controlled rectifier can not feed the power back from dc to ac side while a fully-controlled circuit can do that. This is true for single as well as poly-phase circuits.

In the diagram of input/output voltage waves α is the angle where the thyristors have been fired to conduct. The average dc voltage is given by

$$V_d = \frac{1}{\pi} \int_0^{\alpha+\pi} V_m \sin \omega t \, d\omega t$$

$$= \frac{2}{\pi} V_m \cos \alpha$$

Hence the output voltage can be controlled by controlling the firing angle α.

Table 10.1. Theoretical design data - rectifiers with resistance load

	Single-phase		*Three-phase*		
	Full-wave	*Bridge*	*Half-wave*	*Full-wave*	*Six-phase half-wave*
Value of m	2	2	3	6	6
Average load current, I_{dc}	$0.636V_m$*/R	$0.636V_m/R$	$0.827V_m/R$	$1.65V_m/R$	$0.955V_m/R$
Average current/diode	$I_{dc}/2$**	$I_{dc}/2$	$I_{dc}/3$	$I_{dc}/3$	$I_{dc}/6$
Peak current/diode	$1.571\ I_{dc}$	$1.57\ I_{dc}$	$1.21\ I_{dc}$	$1.21\ I_{dc}$	$1.047\ I_{dc}$
V_{dc} $0.636V_m$	$0.636V_m$	$0.827V_m$	$1.65V_m$	$0.955V_m$	
Lowest ripple frequency	$2f_1$	$2f_1$	$3f_1$	$6f_1$	$6f_1$
Ripple	47%	47%	17%	4%	4%
Transformer V.A. rating :					
Secondary	$1.75P_{dc}$	$1.23P_{dc}$	$1.50P_{dc}$	$1.047P_{dc}$	$1.82P_{dc}$
Primary	$1.23P_{dc}$	$1.23P_{dc}$	$1.23P_{dc}$	$1.047P_{dc}$	$1.28P_{dc}$
Maximum theoretical efficiency	81.2%	81.2%	96.5%	99.8%	99.8%
P.I.V.	$3.14V_{dc}$	$1.57V_{dc}$	$2.09V_{dc}$	$1.045V_{dc}$	$2.09V_{dc}$

* V_m is the peak value of voltage to neutral of one phase of the transformer (see circuit diagrams).
** I_{dc} is defined as the average current in the load.

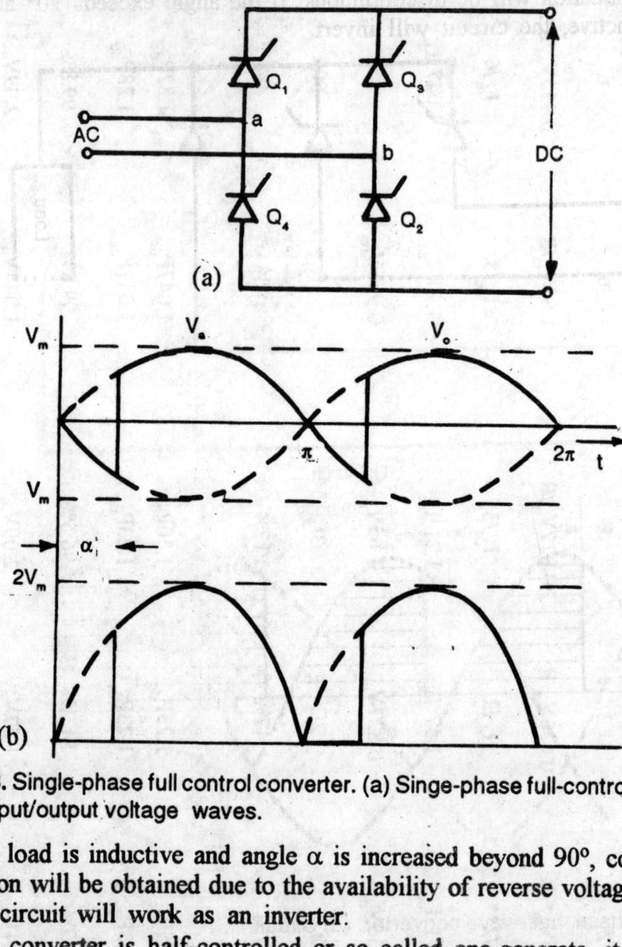

Fig. 10.8. Single-phase full control converter. (a) Singe-phase full-control converter. (b) Input/output voltage waves.

If the load is inductive and angle α is increased beyond 90°, continuous conduction will be obtained due to the availability of reverse voltage at load and the circuit will work as an inverter.

If the converter is half-controlled or so called one-generate, it will not work as inverter.

(ii) Three-phase half-wave converter

The circuit diagram of a 3-phase, half-wave positive converter and the selected wave-shapes are shown in Fig. 10.9. Presence of the input supply neutral is essential for this. The output voltage is given by :

$$V_{dc} = \frac{3}{2\pi} \int_{\frac{\pi}{6}+\alpha}^{\frac{5\pi}{6}+\alpha} V_m \sin \omega t \, d\omega t = 0.675 V_L \cos \alpha$$

where V_m is peak value of phase voltage and V_L is the line rms voltage of angle α is less than 30°, the conduction will be continuous. If it exceeds 30°

then the conduction will be discontinuous. If the angle exceeds 90° and the load is inductive, the circuit will invert.

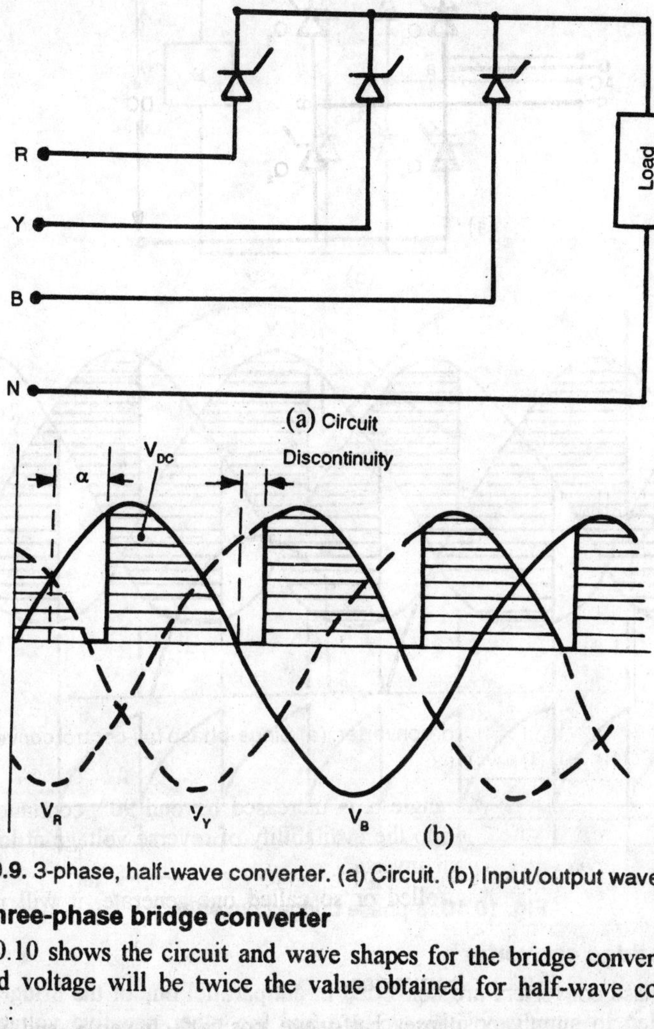

Fig. 10.9. 3-phase, half-wave converter. (a) Circuit. (b) Input/output wave stapes.

(iii) Three-phase bridge converter

Fig. 10.10 shows the circuit and wave shapes for the bridge converter. The dc load voltage will be twice the value obtained for half-wave converter. Hence :

$$V_d = 1.35\ V_L \cos\alpha$$

For angle a more than 90° and less than 180° the circuit will operate in inverting mode.

(iv) Three-phase half-controlled converter

If the lower three or the upper three thyristors of the bridge converter are replaced by diodes, we get the half-controlled converter. The wave shape is shown in Fig. 10.11. The dc load voltage is given by :

$$V_d = 0.675\ V_L\ (1 + \cos\alpha)$$

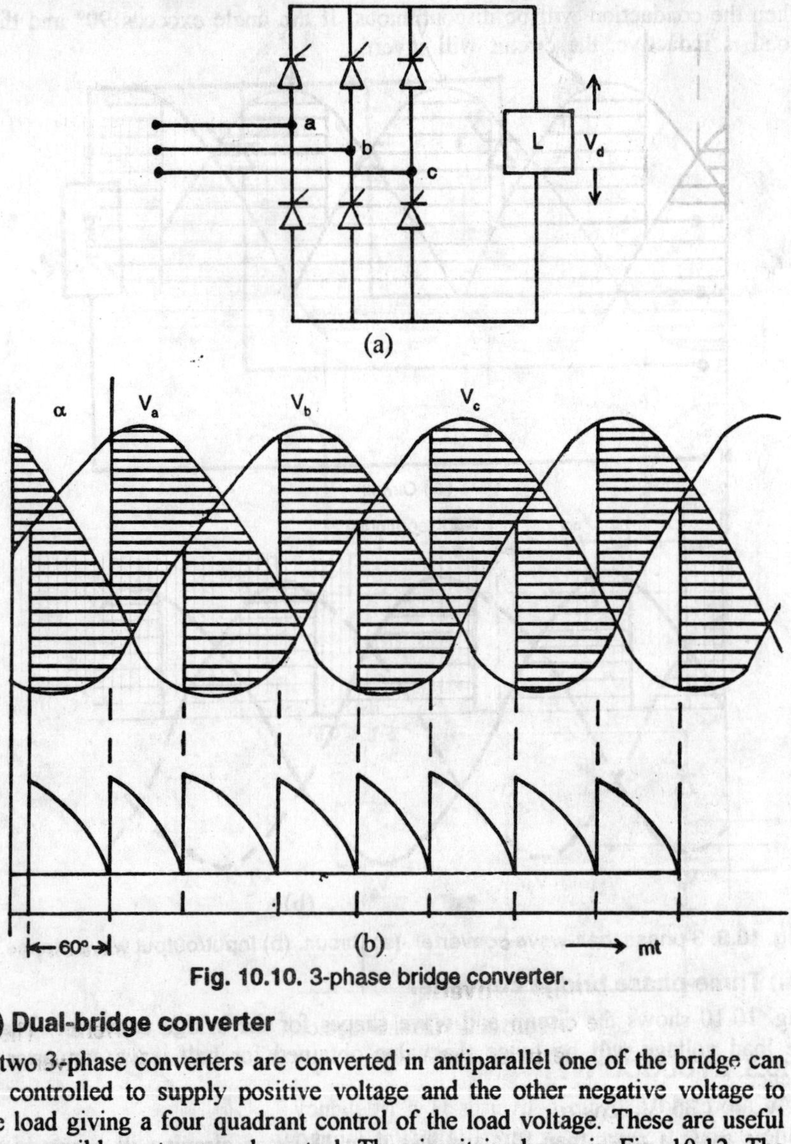

Fig. 10.10. 3-phase bridge converter.

(v) Dual-bridge converter

If two 3-phase converters are converted in antiparallel one of the bridge can be controlled to supply positive voltage and the other negative voltage to the load giving a four quadrant control of the load voltage. These are useful for reversible motor speed control. The circuit is shown in Fig. 10.12. The reactor IGR (intergroup reactor) is provided to have a circulating current between the two groups. In that case one operates as a rectifier and the other as an inverter.

(vi) Multiphase converters

To reduce harmonies in the output wave, 6 or 12 pulse series or parallel converters may be used. The principle of operation remains the same.

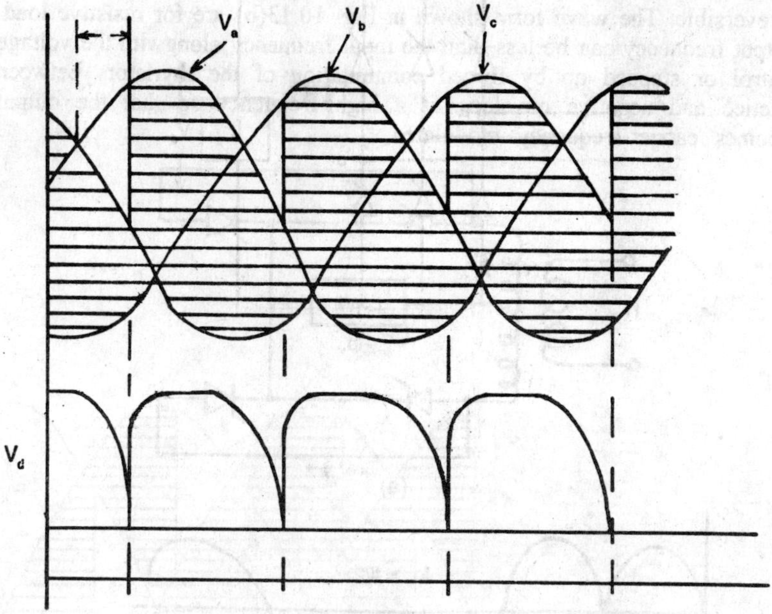

Fig. 10.11

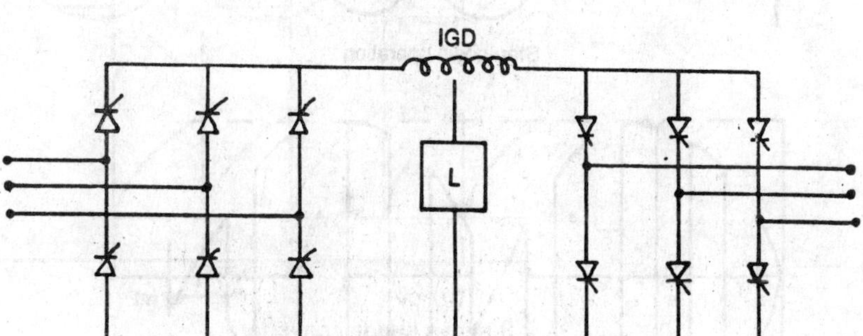

Fig. 10.12. Dual-bridge converter.

10.5. CYCLOCONVERTERS

As aforesaid a cyclo converter is a frequency converter of single-stage. If the output frequency is more than the input frequency, the cycloconverter is called a step-up converter else a step-down cycloconverter. The cycloconverter can also provide a voltage control alongwith frequency control. Various configurations for cycloconverters are possible.

(i) Single-phase/single-phase cycloconverter

This is shown in circuit diagram of Fig. 10.13(a) wherein a positive and a negative centre tap converter are connected in parallel so that load voltage is reversible. The wavè form shown in Fig. 10.13(b) are for resistive load. Output frequency can be less than the input frequency alongwith the voltage control or stepped up by forced commutation of the thyristors between practice and negative envelops at a high frequency so that the output becomes carrier-frequency modulated.

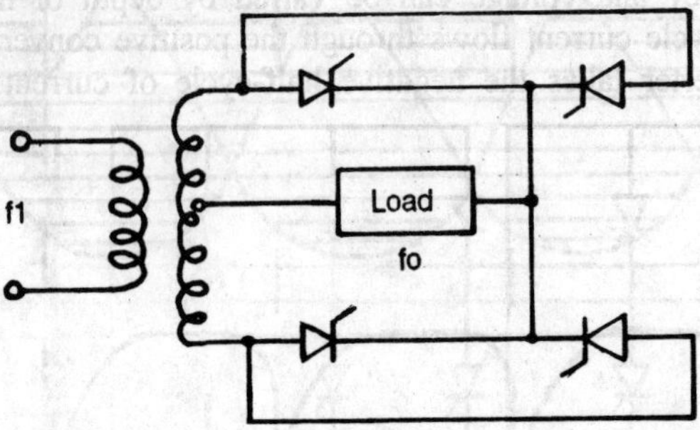

(a)

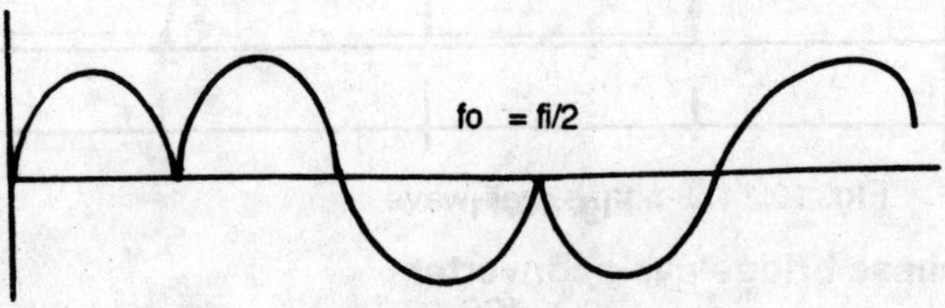

Step-down operation

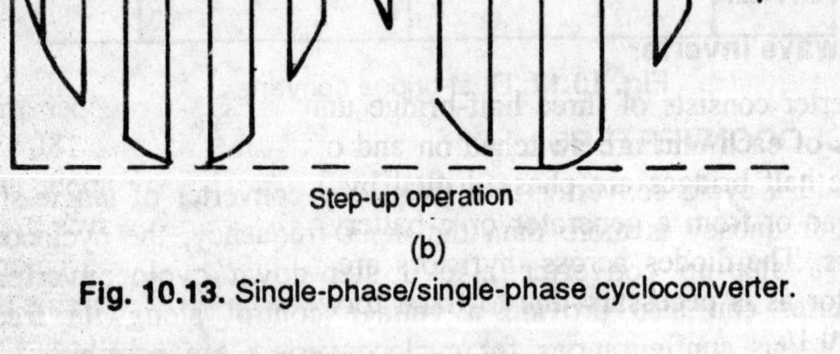

Step-up operation

(b)

Fig. 10.13. Single-phase/single-phase cycloconverter.

(ii) Three-phase half-wave cycloconverter

This circuit is shown in Fig. 10.14. This also known as 18-thyristor, three-pulse cycloconverter. It consists of three identical half-wave anti-parallel phase group. This is equivalent to 3 full-wave converters. If the load neutral is connected to supply neutral, if available, the three phases operate independently. Each phase group functions as a converter with four quadrant capability, hence the load can be supplied with voltage and currents of either polarity. The firing angle of each phase group is modulated sinusoidally but with 110° phase shift so as to fabricate mean sinusoidal output voltage. The output frequency and voltage can be varied by depth of modulation. The positive half cycle current flows through the positive converter whereas the negative converter takes the negative half-cycle of current.

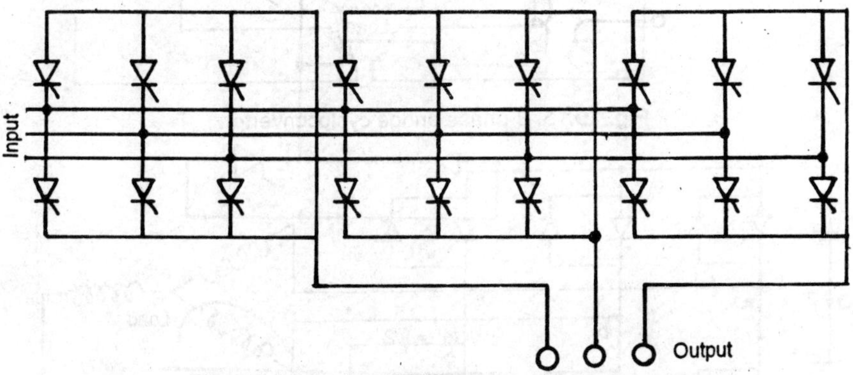

Fig. 10.14. 3-phase, half-wave cycloconverter.

(iii) Three-phase bridge cycloconverter

In this circuit shown in Fig. 10.15 each phase group consists of dual bridge converter with IGR and the load phases are shown isolated. If this type of isolation is not possible in the load, individual phase groups can be isolated at the input using transformers. With bridge connection the output voltage is doubled as compared with half-wave cycloconverter, hence the thyristor current rating is halved, for the same load. This circuit allows smooth output voltage wave fabrication.

10.6. INVERTER

Square wave inverter

This inverter consists of three half-bridge units where the upper and lower thyristors of each unit are switched on and off alternately for 180° interval. The three half bridges are phase drifted by 120°. The dc supply may be a rectified ac or from a generator or a battery. The output waves are rich in harmonics. The diodes across thyristors are helpful in forced commutation of thyristor as is necessary due to fixed forward bias. The circuit is shown in Fig. 10.16.

Taking phase voltages with respect to the centre top of dc supply and considering only fundamental component of voltages :

$$V_{ao} = \frac{4V}{\pi} \cos \omega t$$

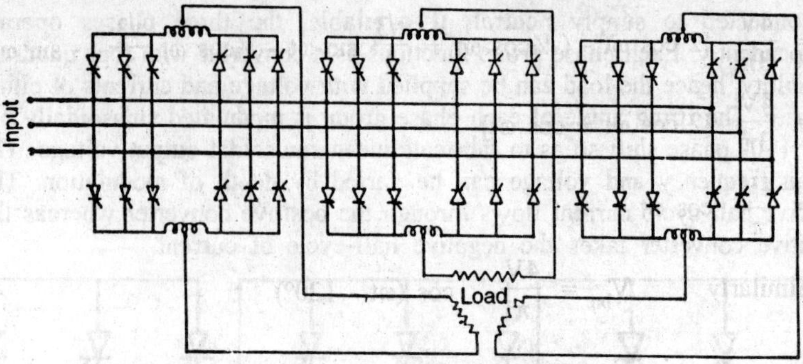

Fig. 10.15. 3-phase bridge cycloconverter.

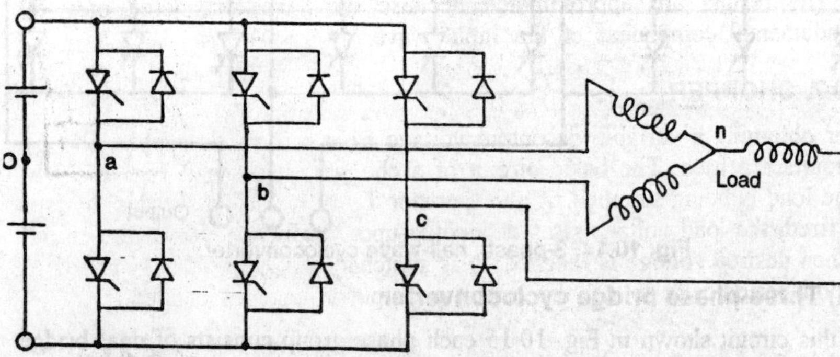

Fig. 10.16. Square wave inverter.

$$V_{bo} = \frac{4V}{\pi} \cos(\omega t - 120^o)$$

$$V_{co} = \frac{4V}{\pi} \cos(\omega t + 120^o)$$

taking $V_{ab} = V_{ao} - V_{bo}$ etc. for three line voltages.

For a balanced supply the neutral voltage will be :

$$V_{no} = (V_{ao} + V_{bo} + V_{co})/3$$

Therefore $V_{an} = V_{ao} - V_{no} = \frac{2}{3} V_{ao} - \frac{1}{3}(V_{bo} + V_{co})$

$$= \frac{2}{3} \frac{4V}{\pi} \cos \omega t - \frac{1}{3} \frac{4V}{\pi} [\cos(\omega t - 120^o) + \cos(\omega t + 120^o)]$$

$$= \frac{4V}{3\pi} [2 \cos \omega t - (\cos \omega t \cos 120^o + \sin \omega t \sin 120^o + \cos \omega t \cos 120^o - \sin \omega t \sin 120^o)]$$

$$= \frac{4V}{3\pi} \left[2 \cos \omega t - \left(-\tfrac{1}{2}\cos \omega t + \frac{\sqrt{3}}{2} \sin \omega t - \tfrac{1}{2}\cos \omega t \frac{\sqrt{3}}{2} - \sin \omega t\right)\right]$$

$$= \frac{4V}{3\pi} [2 \cos \omega t + \cos \omega t]$$

$$= \frac{4V}{\pi} \cos \omega t$$

Similarly $V_{bn} = \frac{4V}{\pi} \cos (\omega t - 120^o)$

$$Vcn = \frac{4V}{\pi} \cos (\omega t + 120^o)$$

The results are approximated because we have considered only the fundamental component of the input wave.

10.7. CHOPPER

For obtaining a variable dc output voltage from a fixed dc supply, chopper circuits are used. The basic circuit of a chopper is given in Fig. 10.17(a). The load is being supplied by the thyristor T_1 on its firing. As this thyristor is fired, the load voltage rises depending upon the inductance of the circuit. When desired voltage is reached T_1 is switched off. Thus T_1 is switched off or on periodically to get the effective output voltage as desired.

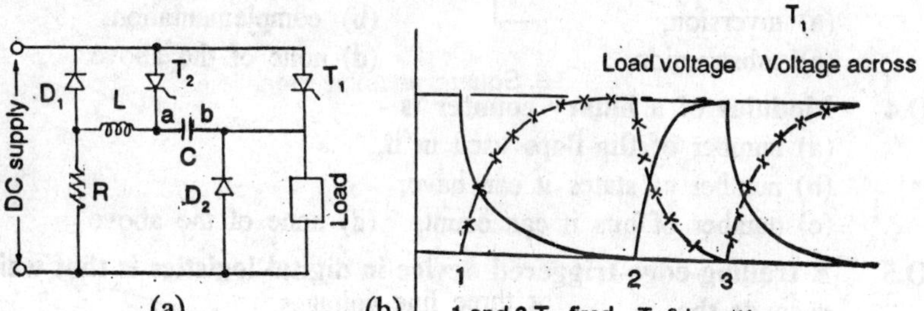

Fig. 10.17. Chopper.

Thyristor T_2, diode D_2 and capacitor C_1 are required to commutate the thyristor T_1. When T_1 conducts, capacitor C_1 is charged with b as positive and a negative. When T_2 is fired, capacitor voltage reverse biases the thyristor T_1 and if its gate pulse is not present, it is switched off. At this the capacitor is reverse charged. When T_1 is again fired to conduct, capacitor voltage reverse-biases the thyristor T_2 and switch it off. Thus rating of thyristor T_2 is just sufficient to change the capacitor.

Voltage wave shapes of the two thyristor and the capacitor are shown in Fig. 10.17(b).

11

Digital Electronics and Computers

Q.1 In sequential logic circuits, some feedback is -
(a) essential, (b) desirable,
(c) not essential, (d) none of the above.

Q.2 In logic circuits, if 1 represents a low voltage and 0, a high voltage, then the logic is called a -
(a) positive logic, (b) negative logic,
(c) binary logic, (d) none of the above.

Q.3 In logic circuits, A bubble means -
(a) inversion, (b) complementation,
(c) substraction, (d) none of the above.

Q.4 Modulus of a binary counter is -
(a) number of flip-flops used in it,
(b) number of states it can have,
(c) number of bits it can count, (d) none of the above.

Q.5 A trailing-edge-triggered device in digital logistics is that which records the -
(a) starting-end of the input pulse,
(b) closing-end of the input pulse,
(c) cloked pulse,
(d) none of the above.

Q.6 In computers, registers are temporary storage of information in binary form. This job is accomplished by the flip-flops. The number of flip-flops in a register is equal to -
(a) half the number of bits of information,
(b) number of bits of infomation,
(c) double the number of bits of information,
(d) none of the above.

Q.7 The number of inputs and outputs of a register are always in the ratio of -

(a) 2 : 1, (b) 1 : 1,
(c) 1 : 2, (d) none of the above.

Q.8 The input to the registered are loaded in it -

(a) by itself, (b) by a clock pulse,
(c) the flipping action, (d) none of the above.

Q.9 All bits of information, in the registers, are (i) parallel-in, parallel-out, (ii) series-in, parallel-out, (iii) series-in, series-out, (iv) parallel-in, series-out.

(a) all the above are true, (b) only first two are true,
(c) only last two are true, (d) only (ii) and (iv) are true.

Q.10 In a bidirectional shift register, the information is shifted from left to right or from right to left, depending upon -

(a) input puse train, (b) a control signal,
(c) unloading of the register, (d) none of the above.

Q.11 Computer memories are of the type (i) semiconductor ICs, (ii) magnetic cores, (iii) magnetic media, (iv) electrostatic charge capacitors.

(a) all the above are true, (b) only first three are true,
(c) only first two are true, (d) none of the above.

Q.12 In reference to computer memory, each input combination of bits, and each output combination of bits, are called -

(a) addresse,
(b) word,
(c) address and word respectively,
(d) none of the above.

Q.13 In an ordinary ROM of a computer, a diodes-matrix is used between the decoder output lines and the memory output lines. The information permanently stored depends upon -

(a) the direction of diode bias,
(b) presence or absence of the diode,
(c) the duration of the input pulse,
(d) none of the above.

Q.14 In a diode-matrix ROM of a computer, the number of diodes present, is equal to -

(a) number of bits of information,
(b) number of 1s in the information,
(c) number of 0s in the information,
(d) none of the above.

Q.15 For programmable-ROMs, the present trend is to store information by the use of-

(a) diodes, (b) BJTs,
(c) MOSFETs (d) none of the above.

Q.16 In a computer ROM the information can be -

(a) stored and retrieved randomly,
(b) stored randomly but retrieved serially,
(c) stored serially but retrieved randomly..

Q.17 In a computer RAM the information can be -

(a) stored and retrieved randomly,
(b) stored randomly but retrieved serially,
(c) stored serially but retrieved randomly.

Q.18 A RAM cell has -

(a) three inputs and one output,
(b) two inputs and two outputs,
(c) ome input and three outputs,
(d) none of the above.

Q.19 For a computer RAM, if address input is n bits and output bits are m, then the RAM is called a -

(a) n x m bit RAM, (b) 2n x m bit RAM,
(c) n x 2m bit RAM, (d) none of the above.

Q.20 In a computer memory, if the information stored, if not altered and power supply is maintained, maintains the status by itself, is known as -

(a) static RAM, (b) RAM,
(c) dynamic RAM, (d) none of the above.

Q.21 A static RAM is composed of -

(a) BJTs only, (b) BJTs, NMOSs or CMOSs,
(c) NMOSs and capacitors, (d) none of the above.

Q.22 A dynamic RAM is composed of -

(a) BJTs only, (b) BJTs, NMOSs or CMOSs,
(c) NMOSs and capacitors, (d) none of the above.

Q.23 In the modern RAMs, the information is stored by charge on the capacitor, which gradually discharges hence refreshing of the memory is cssential. Such a RAM is called -

(a) static RAM, (b) dynamic RAM,
(c) volatile RAM, (d) none of the above.

Q.24 The magnetic core memory, the output is obtained -

(a) by destructing the information content,
(b) without any change in information,

(c) anyway (a) or (b),
(d) by none of the above.

Q.25 The computer floppy disk is a -
(a) magnetic core memory, (b) moving surface memory,
(c) sequential memory, (d) none of the above.

Q.26 Each sector of HD floppy disk has -
(a) 256 bytes, (b) 512 bytes,
(c) variable bytes, (d) none of the above.

Q.27 The quantum of information stored per track on a magnetic floppy disk is -
(a) dependent upon the diameter, (b) fixed,
(c) adjustable, (d) none of the above.

Q.28 In a charge-coupled device memory, a large number of MOSFET gates are serially arranged on a common substrate. The information is stored in the form of electron density below the gates. It is a -
(a) random access device, (b) serial access device,

Q.29 In a computer arithmatic-logic unit (ALU) and the processor are -
(a) two names of the same thing, (b) different entities,
(c) a combination, (d) none of the above.

Q.30 In computers, accumulators are the temporary storages for storing -
(a) data from memory, (b) ALU results,
(c) any data, (d) none of the above.

Q.31 Computer control unit is the combination of -
(a) counters, combinational logic circuits and registers,
(b) counters, logic circuits and registers,
(c) processor, counter and registers,
(d) none of the above.

Q.32 The combination of control unit and the processor is called -
(a) arithmetic logic unit, (b) central processing unit,
(c) motherboard, (d) none of the above.

Q.33 In computers, whole CPU is sometimes made on a single chip. This chip is known as -
(a) processor, (b) microprocessor,
(c) ALU, (d) none of the above.

Q.34 Every particular computer manipulates information in fixed group of bits. This number of bits is called the computer's -

(a) byte size,. (b) word size,
(c) cluster size, (d) none of the above.

Q.35 Modern computeer arithmatic logic units use negative number representation system derived as -
(a) binary system, (b) ones complement system,
(c) twos complement system, (d) none of the above.

Q.36 In twos complement number representation system, a negative number can be converted to a positive number, and vice-versa, by taking -
(a) bitwise complement, (b) bitwise complement plus one,
(c) bit rotation, (d) none of the above.

Q.37 Twos complement (N*) of a number (N) is given by N* = 2n - N, where n is the -
(a) number of bits, (b) square of N,
(c) square root of N, (d) none of the above.

Q.38 In sign and magnitude bit system, the differnce in a positive and negative number is that of -
(a) the first bit, (b) each and every bit,
(c) some bits, (d) none of the above.

Q.39 In twos-complement substraction, carry-out digit is -
(a) added back with end-around, (b) ignored,
(c) retained as such, (d) none of the above.

Q.40 In twos-complement arithmetic operations, overflow generally occur, when -
(a) carry-in and carry-out are equal,
(b) carry-in and carry-out are not equal,
(c) carry-in > carry-out,
(d) none of the above.

Q.41 In a computer, a 32 bit ALU will have -
(a) 32 half-adders, (b) 64 half-adders,
(c) 16 half-adders, (d) none of the above.

Q.42 In a single accumulator computer, the instructions are called single-address instruction. The only address in the instruction is that of -
(a) accumulator, (b) memory location,
(c) operation, (d) none of the above.

Q.43 In a computer an accumulator is used to store -
(a) instructions, (b) data,
(c) memory addresses, (d) none of the above.

Q.44 In a computer, MAR gets the input from -

(a) accumulator, (b) IR,
(c) ALU, (d) none of the above.

Q.45 Harvard computer architecture is considered to be faster than the Princeton architecture becaause the former -

(a) has less steps, (b) can prefetch instructions,
(c) is conceptually simpler, (d) none of the above.

Q.46 In Princeton architecture of computers, data and instructions are stored in -

(a) different memories, (b) same memory,
(c) any of the above, (d) none of the above.

Q.47 In computer engineering, 'register transfer notation' means -

(a) control sequence,
(b) order of registers,
(c) register nomenclature system,
(d) none of the above.

Q.48 in a computer, processor and memory interact -

(a) directly, (b) through MAR and MBR,
(c) through MBR only, (d) none of the above.

Q.49 In the modern computers, with memory-mapped I/O strategy, processor and I/O devices are connected through -

(a) a separate dedicated bus, (b) the processor-memory bus,
(c) through memory, (d) none of the above.

Q.50 In modern computers, with memory-mapped I/O strategty, all I/O devices are treated as -

(a) memory with different addresses,
(b) components of main memory,
(c) as processor parts,
(d) none of the above.

Q.51 In modern computers, the data transfer between the main memory and the I/O devices is -

(a) direct, (b) through processor,
(c) through dedicated bus, (d) none of the above.

Q.52 In computer systems, polling and interrupts are -

(a) the names of the same action,
(b) the actions for the same purpose,
(c) the actions for different purposes,
(d) none of the above.

Q.53 Polling in computer systems, is an action of I/O devices, which means, on completion of a given operation, it -

(a) places its status in the register,
(b) sends control signal to the processor,
(c) forces the processor to provide it more input,
(d) none of the above.

Q.54 Interrupt in computer systems, is an action of I/O devices, which means, on completion of a given operation, it -
(a) places its status in the register,
(b) sends control signal to the processor,
(c) forces the processor to provide it more input,
(d) none of the above.

Q.55 In most of the general purpose computers, for informing the processor of a completed action, I/O devices use -
(a) interrupts, (b) polling,
(c) both interrupts and polling, (d) none of the above.

Q.56 In computers the processors efficiency is defined as the number of control states (or clock cycles) needed to -
(a) execute a typical instruction,
(b) fetch a typical instruction,
(c) fetch and execute a typical instruction,
(d) none of the above.

Q.57 For computers, the most flexible bus arrangement is a point to point connection but it is rarely used, because of the -
(a) low resource utilization, (b) high hardware cost,
(c) low system efficiency, (d) none of the above reasons.

Q.58 In computers a swap operation means transfer of data -
(a) from one register to another,
(b) from one bus to another,
(c) between any two components,
(d) none of the above.

Q.59 In computers, each input/output device is driven by its own computer because of their low speed operations, to save on -
(a) total computer cost, (b) processor time,
(c) efficiency, (d) none of the above.

Q.60 At the start of a computer, it is desirable to set the PC to zero and drive the memory request line to unasserted value. This process is known as -
(a) booting, (b) starting,
(c) resetting, (d) none of the above.

Q.61 Tokenization means providing values to each element of a program. This job is done by -

(a) application software, (b) system program,
(c) compiler, (d) none of the above.

Q.62 An imperative language is that in which assigns definite values are provided to each program variable, Examples of such languages are -

(a) C and BASIC, (b) FORTRAN and COBOL,
(c) none of the above.

Q.63 For scientific and engineering applicaions, FORTRAN is more appropriate because it is an -

(a) imperative language, (b) applicative language,
(c) none of the above.

Q.64 A computer can be made to behave like another computer through software simulation, such a computer is called -

(a) hardware computer, (b) ghost computer,
(c) software-simulated computer, (d) none of the above.

Q.65 The attribute of a computer language of being able to combine various features of the language in all possible combinations, is called -

(a) naturalness, (b) abstractability,
(c) orthogonality, (d) none of the above.

Q.66 Certain words are chosen by a computer language for specified purposes, these can not be used by the programmer for other purposes. Such special words are called -

(a) tokens, (b) noise words,
(c) key words or reserved words, (d) delimiters,

Q.67 Certain computer languages use optional words to improve program readability, such words are called -

(a) tokens, (b) noise words,
(c) key words or reserved words, (d) delimiters,

Q.68 The syntactic elements such as paranthesis and inverted commas are used by programming languages to remove ambiguities of statements. Such elements are known as -

(a) tokens, (b) noise words,
(c) key words or reserved words, (d) delimiters,

Q.69 A microcomputer is that computer, which is -

(a) the currently smallest-sized computer,
(b) based upon a microprocessor,
(c) accomodable on a table-top, (d) none of the above.

Q.70 At a timesharing computer system, various user of the computer use the main facility -

(a) one at a time,
(b) simultaneously through their teerminals,
(c) none of the above.

Q.71 A multitasking computer system is that which -

(a) can be used by various users simultaneously,
(b) performing multiple operations simultaneously,
(c) can be used for multiple purposes,
(d) none of the above.

Q.72 In a distributed processing comuter system also known as multiprocessing, the main computer facility is available to various users through -

(a) terminals at their desks,
(b) smaler computerts at their desks,
(c) none of the above.

Q.73 In a microcomputer if there are 16 liones on the adress-bus, at a time, the CPU can address nearly -

(a) 16,000 memory locations, (b) 64,000 memory locations,
(c) 1,28,000 memory locations, (d) none of the above.

Q.74 In a microcomputer, all devices connected to a data bus have -

(a) two-state (binary) output, (b) three-state output,
(c) multi-state output, (d) none of the above.

Q.75 The main difference between the microprocessors known as 8088 and 8086 is -

(a) in word sizes of 8-bit and 16-bit respectively,
(b) in word sizes of 16-bit and 8-bit respectively,
(c) in size of the address bus,
(d) none of the above.

Q.76 Pipelining in a microprocessor means fetching next instruction,

(a) immediately after the current task is executed,
(b) during the execution of the current task,
(c) buffering the gap betwwen the instructions,
(d) none of the above.

Q.77 The 8086 microprocessor, although sends out 20-bit addreses and thus can address any of the 1 MB addreses in memory, at a given time, it works with only 4 segments each of -

(a) 16 KB (b) 64 KB,
(c) 128 KB, (d) 250 KB.

Q.78 The four segment registers in the BIU of 8086 microprocessors, contain -
(a) starting addresses of the four working segments of memory,
(b) sizes of the four working segments of the memory,
(c) operations to be performed on the four segments,
(d) none of the above.

Q.79 The instruction pointer register of the 8086 microprocessor holds the 16 bits of the offset to be added to the contents of the -
(a) code segment register, (b) stack segment register,
(c) data segment register, (d) none of the above.

Q.80 The stack in a computer system is a section of memory to hold addresses and data while the processor is busy in execution. The stack -
(a) is always predefined,
(b) is randomly decided by the processor,
(c) size varies with processor execution,
(d) none of the above.

Q.81 All computer systems processors, basically get input and provide output in a language that can be -
(a) machine language only, (b) assembly language,
(c) any high-level language, (d) none of the above.

Q.82 In a computer system an asembler is a program that translates-
(a) high-level language into asembly language,
(b) assembly language into machine language,
(c) high-level language into machine language,
(d) none of the above.

Q.83 8086 microprocessor of computer systems keeps first 1 KB of memory reserved for interrupts service procedures, wherein it can store -
(a) 64 interrupts,
(b) starting addresses of 128 interrupt procedures,
(c) starting addresses of 256 interrupt procedures,
(d) none of the above.

Q.84 An interrupt vector in the 8086 microprocessor interrupt-pointer table is of the size of -
(a) one byte, (b) one word,
(c) two words, (d) none of the above.

Q.85 In computer engineering handshaking means -
(a) mutual transfer of data,
(b) receiving data and acknowledging,

(c) receiving data-ready signal and acknowledging,
(d) none of the above.

Q.86 In computer systems, while transmitting data at high speeds, alongwith the data the sender also sends a signal to the receiver to inform him the availability of data on line. This signal is known as -

(a) handshake, (b) interrupt,
(c) strobe, (d) none of the above.

Q.87 In direct memory access of a microcomputer, the data is directly flown between the peripheral device and the memory, without the use of micoprocessor. With this facility -

(a) microprocessor and the DMA timeshare the use of buses,
(b) DMA and the memory timeshare the use of microprocessor,
(c) buses and memory timeshare the use of microprocessor,
(d) none of thc above.

Q.88 In computers, large memory devices, known as dynamic RAMs, store information bits in -

(a) flip-flops, (b) capacitors,
(c) inductors, (d) none of the above.

Q.89 In dynamic RAM system in microcomputers, refreshing of the memory at regular intervals is essential, because -

(a) memory flip-flops gets disturbed,
(b) memory capacitors get discharged,
(c) microprocessor changes the stste,
(d) none of the above.

Q.90 The refreshing operatiqn of dynamic RAM in a microcomputer is require every -

(a) 1 msecond, (b) 2 to 8 mseconds,
(c) every 80 to 100 mseconds (d) eveery 1 minute.

Q.91 Read/write operation and refreshing operation on a dynamic RAM in a microcomputer -

(a) can be carried out simultaneously,
(b) can't be carried out simultaneously,
(c) have no mutual relativity,
(d) none of the above.

Q.92 Cache memory in a microcomputcr is -

(a) low-capacity high-speed RAM,
(b) low-speed high-capacity RAM,
(c) high-speed high-capacity RAM,
(d) none of the above.

Q.93 In a microcomputer, other memory other than the main RAM, which can be randomely accessed is -

(a) ROM,

(b) none,

(c) cache,

(d) all nonmagnetic memories.

Q.94 What is the binary number of decimal numbers 0.25 and 2.5 ?

(a) 0.11, 10.1, (b) 0.11, 1.1,

(c) 1.01, 10.1, (d) none of the above.

Q.95 What is the octal number of decimal number 8.8 ?

(a) 10.6314, (b) 1.1,

(c) 1.6314, (d) none of the above.

Q.96 What is the binary number of the octal number 4.1 ?

(a) 100.001, (b) 110.111,

(c) 100.100, (d) none of the above.

Q.97 The digits of a hexadecimal number system are -

(a) 0,1,2,3,4,5.

(b) 0,1,2,3,4,5,6,7,8,9,10,11,12,13,14,15.

(c) 0,1,2,3,4,5,6,7,8,9,A,B,C,D,E,F..

(d) none of the above.

Q.98 The binary-coded decimal of decimal (BCD) number 12 is -

(a) 0001 0010 (b) 1100

(c) 12 (d) none of the above.

Q.99 The usual staircase light switching circuit represents -

(a) an AND logic, (b) an OR logic,

(c) an exclusive-OR logic, (d) none of the above.

Q.100 The relationship L = A(B + C) = AB + AC, is known as -

(a) additive law, (b) distributive law,

(c) multiplication law, (d) none of the above.

Q.101 The truth table -

Input1	Input2	Output
0	0	0
0	1	1
1	0	1
1	1	0

represents -

(a) OR gate, (b) NOR gate,

(c) exclusive-OR gate, (d) none of the above.

Q.102 The truth table -

Input1	Input2	Output
0	0	1
0	1	1
1	0	1
1	1	0

represents -

(a) OR gate, (b) NAND gate,
(c) exclusive-OR gate, (d) none of the above.

Q.103 The truth table -

Input1	Input2	Output
0	0	1
0	1	0
1	0	0
1	1	0

represents -

(a) OR gate, (b) NOR gate,
(c) exclusive-OR gate, (d) none of the above.

Q.104 In Boolean algebra, inversion of inversion is called -

(a) evolution, (b) involution,
(c) double inversion, (d) none of the above.

Q.105 In boolean algebra, relationships AB = BA and (A + B) = (B + A) are termed as -

(a) associative law, (b) distributive law,
(c) commutative law, (d) none of the above.

Q.106 In boolean algebra, relationships A(BC) = (AB)C and (A + B) + C = A + (B + C) are termed as -

(a) associative law, (b) distributive law,
(c) commutative law, (d) none of the above.

Q.107 According to the absorption laws of Boolean algebra values of (A + AB) and A(A + B) are respectively -

(a) A and A, (b) A and B,
(c) B and A, (d) none of the above.

Q.108 If iA represents invert-A, then according to the absorption laws of Boolean algebra values of (A + iAB) and A(iA + B) are respectively -

(a) A and A, (b) A and B,
(c) (A+B) and AB, (d) none of the above.

Q.109 If iA represents invert-A, according to De Morgan's theorem of Boolean algebra values of i(A+B) and i(AB) are respectively -

(a) (iA.iB) and (iA+iB), (b) (iA+iB) and (iA.iB),
(c) (AB) and (A+B), (d) none of the above.

Q.110 In Boolean algebra, in an expression, if all the 1s are changed to 0s and 0s to 1s, and all the AND and OR operations are interchanged, the resultant expression is said to be the -
(a) invert of original expression,
(b) dual of original expression,
(c) complement of original expression,
(d) none of the above.

Q.111 If D(A) represents dual of A, and iA represents invert of A, then for equation A = B, we have -
(a) D(A) = D(B), (b) D(A) = D(iB),
(c) D(A) = -D(B), (d) none of the above.

Q.112 All possible products, containing the factors equal to the number of variables, of either the variables or their complements of a Boolean function, are called -
(a) expressions, (b) maxterms,
(c) minterms, (d) none of the above.

Q.113 Logically a NAND gate can be realized from a NOT (invert) gate.
(a) true, (b) false,
(c) partially true, (d) nonsense.

Q.114 All Boolean functions can be realized by two single types of gates, known also as universal gates. These gates are of the types -
(a) OR gate and NOR gate,
(b) AND gate and NAND gate,
(c) NAND gate and NOR gate,
(d) none of the above.

Q.115 All possible sums, with components equal to the number of variables, of either the variables or their complements, of a Boolean function, are called -
(a) expressions, (b) maxterms,
(c) minterms, (d) none of the above.

Q.116 Standard Boolean function means (i) sum of the minterms, (ii) product of maxterms, (iii) any combination of minterms and maxterms.
(a) all the above are true, (b) only first two are true,
(c) only last two are true, (d) only (i) and (iii) are true.

Q.117 Basic gates of IC logic are -
(a) AND and OR gates, (b) NAND and NOR gates,
(c) AND and NAND gates, (d) OR and NOR gates.

Q.118 A boolean function is called to be in canonical form, if it is expressed as -

(a) sum of minterms, (b) product of maxterms,
(c) either (a) or (b), (d) (a) and (b).

Q.119 A sequential logic output not only depends upon the input to the system but also upon -

(a) state of the system, (b) temperature ofthe system,
(c) other external factors, (d) none of the above.

Q.120 In logic circuits, the set-reset function is provided by -

(a) gates, (b) filters
(c) flip-flop, (d) none of the above.

Q.121 A multivibrator circuit has two outputs, available alternately, by one of the two inputs, and depending upon -

(a) the input, (b) last output,
(c) present circuit state, (d) none of the above.

Q.122 A bistable multivibrator, is so called because -

(a) it oscillates between two states,
(b) it can stay in any one of the two states,
(c) it returns to a predetermined state.
(d) none of the above.

Q.123 A monostable multivibrator, is so called because -

(a) it oscillates between two states,
(b) it can stay in any one of the two states,
(c) it returns to a predetermined state..
(d) none of the above.

Q.124 Clocking or gating of a flip-flop is to accept the input only at predetermined state. It is achieved by the use of AND gates, which get the input alongwith -

(a) its complement, (b) a clock pulse,
(c) enabling pulse, (d) none of the above.

Q.125 In digital logistics, a synchronous sequential circuit is that which accepts inputs subject to -

(a) complementation,
(b) fixed speed,
(c) the presence of a clock pulse,
(d) none of the above.

Q.126 In a data or delay flip-flop, both the inputs terminals always get inputs complemented to each other. In such a logic device the outputs, also of complemetary nature follow the input. The advantage of this arrangement is that the flip-flop -

(a) always get clocked pulses,
(b) is state is altered with each pulse,
(c) never gets both inputs equal to one,
(d) none of the above.

Q.127 In a J-K flip-flop, additional provisions are provided to gain the special feature of -
(a) working at faster speed, (b) auto-reset,
(c) accepting all types of inputs, (d) none of the above.

Q.128 In a toggle flip-flop, both the inputs of a J-K flip-flop are joined together. By this feature we get, what is known as -
(a) data (D) flip-flop, (b) delay flip-flop,
(c) S-R flip-flop, (d) toggle (T) flip-flop..

Q.129 The main drawback of the toggle (T) flip-flop is that if the incoming pulse duration is extended, the circuit goes on changing states. Retaining its advantageous features, the drawback has been overcome in the -
(a) J-K flip-flop, (b) data flip-flop,
(c) JK master-slave flip-flop, (d) none of the above.

Q.130 In digital logistics, an excitation table shows -
(a) inputs and effects, (b) results and causes,
(c) inputs and outputs, (d) none of the above.

Q.131 A flip-flop works as a -
(a) sequential logic, (b) combinational logic,
(c) both the above, (d) none of the above.

Q.132 In a digital multiplexer (mux), there are a number of inputs and a single output. The selector lines make the output equal to -
(a) sum of the inputs,
(b) 0 or 1 depending upon the number of inputs,
(c) one of the inputs,
(d) none of the above.

Q.133 In a digital demultiplexer (demux), there are -
(a) multiple inputs and multiple outputs,
(b) multiple inputs and single output,
(c) single output and multiple inputs,
(d) single input and single output.

Q.134 An encoder converts -
(a) information to binary codes,
(b) binary codes to information,
(c) any way (a) or (b),
(d) none of the above.

Q.135 In digital logistics, if any entry can be freely choosen to be a 0 or a 1, then the entry is designated as -

(a) impotent, (b) redundant,
(c) don't care, (d) none of the above.

Q.136 In logic circuits, if 0 represents a low voltage and 1, a high voltage, then the logic is called a -

(a) positive logic, (b) negative logic,
(c) binary logic, (d) none of the above.

Q.137 In logic circuits, if 1 represents a low voltage and 0, a high voltage, then the logic is called a -

(a) positive logic, (b) negative logic,
(c) binary logic, (d) none of the above.

Q.138 The truth table -

Input1	Input2	Output
0	0	0
0	1	1
1	0	1
1	1	1

represents an (i) AND gate in negative logic, (ii) OR gate in positive logic.

(a) (i) and (ii) are wrong, (b) (i) and (ii) are right,
(c) only (i) is right, (d) only (ii) is right.

Q.139 Modulus of a binary counter is -

(a) number of flip-flops used in it,
(b) number of states it can have,
(c) number of bits it can count,
(d) none of the above.

Q.140 A trailing-edge-triggered device in digital logistics is that which records the -

(a) starting-end of the input pulse,
(b) closing-end of the input pulse,
(c) cloked pulse,
(d) none of the above.

Q.141 In digiital logistics 'decomposition' means replacing a function by a few expressions -

(a) as factors, (b) with fewer literals,
(c) with smaller minterms, (d) none of the above.

Q.142 In digital logistics, 'extraction' means factoring a collection of functions and replacing the common factors by other expressions in order to replace the original set of functions with a new set of functions. In implementation, it reduces the -

(a) number of levels, (b) delay in operation,

(c) number of gates, (d) none of the above.

Q.143 In digital logistics, factoring means replacing a lower level expression of a function with a higher level expression, in order to -

(a) extract common factors,
(b) reduce operational delays,
(c) expand the range of operation,
(d) none of the above.

Q.144 In digital logistics, substitution means replacement of a part of a function by its equivalent, in order to implement -

(a) extraction, (b) factoring,
(c) collapsing, (d) none of the above.

Q.145 In digital logistics, collapsing is reverse of substitution. It is done to -

(a) reduce operational delays, (b) increase number of levels,
(c) decrease number of gates, (d) none of the above.

Q.146 In digital logistics, static-1 hazard means a deviation in expected value of output from -

(a) one to zero without change in input,
(b) zero to one without change in input,
(c) one or zero to zero or one,
(d) none of the above.

Q.147 In digital logistics, dynamic hazard means more-than-one transitions in output when -

(a) input is steady, (b) input is in transition,
(c) circuit is in operation, (d) none of the above.

Q.148 In digital logistics, chances of dynamic hazards increase with incresa in -

(a) number of gates, (b) number of levels,
(c) number of inputs, (d) none of the above.

Q.149 In digital logistics, both PLAs and PALs contain arrays of AND and OR gates, but differ in that (i) in PLAs, both AND and OR arrays are programmable, (ii) in PALs only AND gates are programmable while OR gates are hard-wired.

(a) all the above are true, (b) both the above are false,
(c) only (i) is true, (d) only (II) is true,

Q.150 Due to hard-wired connections in PALs of digital logistics, as compared to the PLAs, they -

(a) are slower in operation, (b) are faster in operation,
(c) have the same speed, (d) none of the above.

Q.151 In digital logistics, gray code means a serialized binary code in which -

(a) alternate digits are respectively 0 and 1,
(b) all digits are either 0 or 1,
(c) only one digit changes at a time,
(d) none of the above.

Q.152 In digital logistics, steering logic circuits are those which route the data inputs to outputs based upon the -

(a) nature of input,
(b) naturen of control signal,
(c) nature of the input and the control signal,
(d) none of the above.

ABSTRACT

11.1. DIGITAL LOGIC

11.1.1. Number system

A number, in general, can be represented by :

$$a_n\ a_{n-1}\ a_{n-2}\ .\ .\ .\ a_3a_2a_1a_o.a_{-1}a_{-2}a_{-3}\ .\ .\ .\ a_{-m}$$

where n represents the number of integral digits and m, the decimal places. The value of this number is given by :

$$a_n \times r^n + a_{n-1}\ r^{n-1} + \ .\ .\ .\ + a_3\ r^3 + a_2r^3 + a_2r^2 + a_1r^1 + a_or^o$$
$$+ a_{-1}r^{-1} + a_{-2}r^{-2} + \ .\ .\ .\ + a_{-m}r^{-m}$$

where r is the radix or base of the number system. Thus for a decimal number with radix 10, this number represents :

$$a_n10^n + a_{n-1}10^{n-1} + \ .\ .\ .\ + a_3 10^3 + a_2 10^2 + a_1 10^1 + a_o 10^o$$

The radix is the base of a number system. We can have any number system by fixing the radix. The number of digits in the number system is equal to the value of radix. Four number systems are in practical use in digital logics. These systems, their radix and the digits are given below :

1. *Binary numbers* : Radix : 2, Digits : 0, 1.
2. *Octal numbers* : Radix : 8, Digits : 0, 1, 2, 3, 4, 5, 7.
3. *Decimal numbers* : Radix : 10, Digits : 0, 1, 2, 3, 4, 5, 6, 7, 8, 9.
4. *Hexadecimal number* : Radix : 16, Digits : 0, 1, 2, 3, 4, 5, 6, 7, 8, 9, A, B, C, D, E, F.

Binary numbers are used in digital computers to represent the presence or absence of a signal. Octal numbers are rarely used. Decimal numbers are used by all of us in daily routine. Hexadecimal numbers are used as shorthand tools for representing binary numbers in explanations.

Binary, octal and hexadecimal numbers are closely related, hence their conversion from are system to the other is easier.

For conversion a binary number of octal, form groups of 3 digits each starting from the binary point to the left and to the right. Convert each group to an octal equivalent, the equivalent digits will give the octal number. To reverse the process, replace each digit of an octal number by three digits of binary number to get the binary equivalent of the whole number.

Similarly, for conversion from hexadecimal to binary and vice versa binary number is used in groups of 4 instead of 3 as in octal numbers.

Complements of binary numbers

(a) 1's complement of a binary number is obtained by replacing all zeros by ones and all ones by zeros.

(b) 2's complement of a binary number is obtained by substracting the number from 2^p.

(c) 2's complement of a binary number can be obtained by adding 1 to the 1's complement of the binary number.

Binary arithmetic rules

1. Addition of binary numbers is done by using usual carries and following the rules :
   0 + 0 = 0, 0 + 1 = 1, 1 + 1 = 10
2. Substraction of binary numbers is done by using usual borrows and the rules :
   0 - 0 = 0, 10 - 1 = 1, 1 - 1 = 0, 1 - 0 = 1
3. Multiplications of binary numbers can be carried out as usual using the rules :
   0 x 0 = 0, 0 × 1 = 0, 1 × 1 = 1
4. Division of binary numbers is done as usual by repeated substractions.

Use of complement numbers

A. The difference between to positive binary numbers n and m can be obtained using the 1's complement as follows :
   1. To n add the 1's complement of m.
   2. If the sum has a carry at the left-end, add 1 to the least significant digit of the sum, the result will be (n -m).
   3. If the sum has no carry at the left-end, take the 1's complement of the sum and place a negative sign before it. The result is (n - m).

B. The difference between two positive binary numbers n and m can be obtained by using 2's complement as follows :
   1. To n add 2's complement of m.
   2. If the sum has a carry at the left-end, remove it, the result is (n - m).
   3. If the sum has no carry at the left-end, take the 2's complement of the sum and put a negative sign before it. The result is (n - m).

Binary-coded-decimal code (BCD code)

When each digit of a decimal number is represented by its binary code of four digits each, the resultant binary number is called a BCD code.

11.1.2. Digital logic operations

(i) OR operation

The result L = A or B, written as (A +B), if given by the following truth table, the operation between A and B is called an OR operation. Physically it can be achieved by the two switched A and B connected in parallel and controlling a common load L. The gate representing this logic is shown in Fig. 11.1(a).

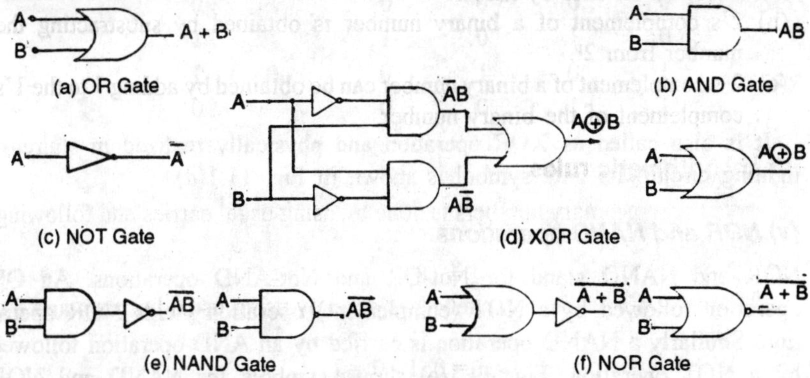

Fig. 11.1. Logic gates.

OR truth table	A	B	L = A + B
	0	0	0
	0	1	1
	1	0	1
	1	1	1

(ii) AND operation

The AND operation between A and B, denoted by AB or Å.B, is defined by the following truth table. It is physically achievable by the two switches connected in series and controlling a common load. The gate representing this logic is shown in Fig. 11.1(b)

AND truth table	A	B	L = AB
	0	0	0
	0	1	0
	1	0	0
	1	1	1

(iii) Complement operation

If A = 0, when B = 1 and A = 1 when B = 0, then A and B are said to be the complements of each other and denoted by :

$A = \overline{B}$ or $B = \overline{A}$

Its gate symbol is shown in Fig. 11.1(c), it is also called a NOT operation.

(iv) Exclusive-OR operation

A	B	AB	AB	AB + AB = A B
0	0	0	0	0
0	1	0	1	1
1	0	1	0	1
1	1	0	0	0

It is also called as XOR operation and physically realized in staircase lighting circuit. Its gate symbol is shown in Fig. 11.1(d).

(v) NOR and NAND operations

NOR and NAND stand for Not-OR and Not-AND operations. An OR operation followed by a NOT (complement) operation yields NOR operation. Similarly a NAND operation is carried by an AND operation followed by a NOT operation. Fig. 11.1(e) shows symbols for NAND and NOR operations. An arbitrary Bolean function can be expressed in terms of NAND gates only, hence NAND gate is known as a universal gate.

11.1.3. Boolean algebra rules

A.1 = A	A + 1 = 1
A.A = A	A + A = A
A.O = 0	A + 0. = A
$A.\bar{A} = 0$	$A + \bar{A} = 1$
Involution	$\bar{\bar{A}} = A$
Commutative law	A.B = B.A. and A + B = B + A
Associative law	A(BC) = (AB)C and A + (B + C) = (A + B) + C
Distributive law	A (B + C) = AB + AC
	A + BC = (A + B) (A + C)
Absorption law	A + AB = A and A (A + B) = A
	A + AB = A + B and A (A + B) = AB
De Morgan's theorem	A + B = AB and AB = A + B

11.1.4. Standard boolean function forms

The canonical forms of Boolean functions are :

(i) The sum of products

Minterms - group of combinations of all variables each being complemented or uncomplemented and joined together by OR operators. All minterms, with value 1, are joined together by AND operators. The result is said to be the sum of products.

For example three variables A, B and C have the sum of products as :

$$F = \bar{A}\bar{B}\bar{C} + \bar{A}B\bar{C} + \bar{A}BC + A\bar{B}C + AB\bar{C} + ABC + ABC + ABC$$

$$= 0 + 1 + 1 + 0 + 1 + 0 + 0 + 1$$

$$= \bar{A}\bar{B}C + \bar{A}B\bar{C} + A\bar{B}\bar{C} + ABC$$

(ii) The product of sums

Maxterms - group of combinations of all variables each being complemented or uncomplemented and joined together by AND operators. All maxterms, with value zero, are joined together by OR operators. The result is said to be the product of sums. For example of three variables A, B and C have the product of sums as :

$$F=(A+B+C)(A+B+\bar{C})(A+\bar{B}+C)(A+\bar{B}+\bar{C})(\bar{A}+B+C)(\bar{A}+B+\bar{C})(\bar{A}+\bar{B}+C)(\bar{A}+\bar{B}+\bar{C})$$

$$= 0 \quad 1 \quad 1 \quad 0 \quad 1 \quad 0 \quad 0 \quad 1$$

$$= (A+B+C) \quad (A+\bar{B}+\bar{C}) \quad (A+\bar{B}+\bar{C}) \quad (\bar{A}+\bar{B}+C)$$

Simplification of bollean functions

Karnaugh maps are used for simplification as follows :

1. Given a boolean function, express it in terms of minterms and place is the corresponding boxes of a Karnaugh map.
2. Group adjascent is so as to form largest possible groups. One or more is can be in more than one groups.
3. A single 1 forms its own group.
4. Take the sum of the terms corresponding to each group, which is the simplified function.

In case maxterms are mapped on the Karnaugh map, zeros are grouped together.

11.2. COMBINATION LOGIC DESIGNS

A combination logic is that which provides output determined by the inputs at that particular moment.

(i) Adders

In addition, the ordinary sum is represented by an exclusive-OR operation while the carry result is given byan AND operation. Logic circuit providing two outputs - one for summation and another for carries for the two inputs, is called a half-adder. It is shown in Fig. 11.2(a).

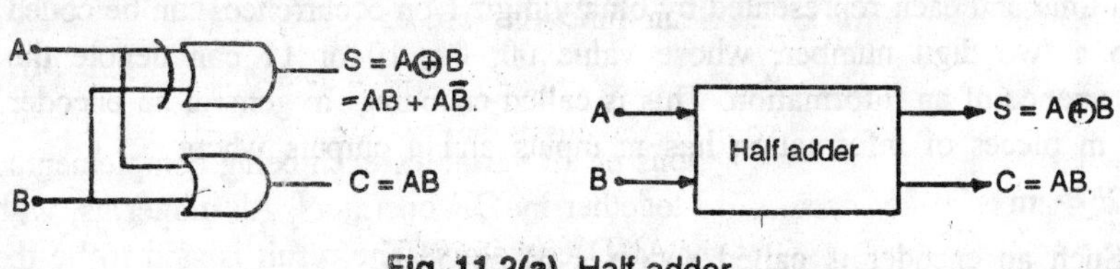

Fig. 11.2(a). Half adder.

The addition operations of sum and carry, can also be expressed in terms of minterms as follows :

$$S = \bar{A}\bar{B}C + \bar{A}B\bar{C} + A\bar{B}\bar{C} + ABC \text{ and } C = AB + BC + CA$$

The circuit representations of adding operations shown separately for sums and carries are known as full adder, shown in Fig. 11.2(b).

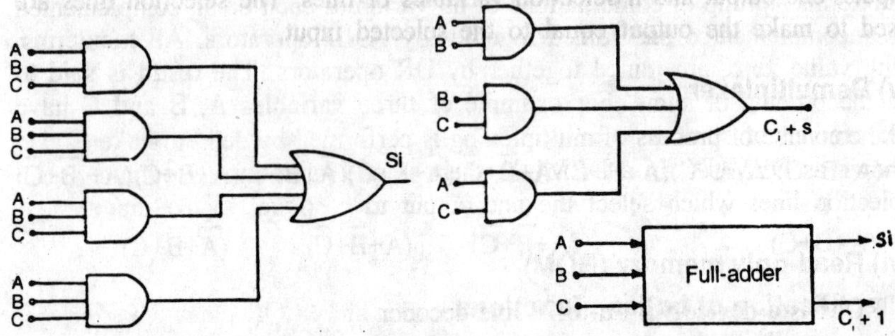

Fig. 11.2(b). Full adder.

By combining one full-adder for each digit and adding the carry in the next digit operation, complete operation of addition is performed as shown in Fig. 11.3.

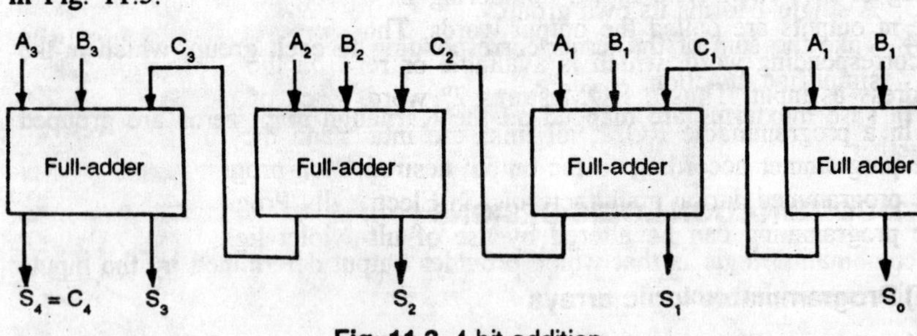

Fig. 11.3. 4-bit addition.

(ii) Encoders

A number of informations, available only one at a time, can be represented by digits of a number. For example four pieces of information, one occurring at a time and each represented by binary digit 1 on occurrence, can be coded into a two digit number, whose value 00, 01, 10 or 11 can denote the occurrence of an information. This is called encoding. In general an encoder for m pieces of information has m inputs and n outputs where

$$2^n > m$$

Such an encoder is called an (m × n) encoder.

(iii) Decoders

The converse of the process of encoding is decoding. In general, an n-to-m decoder has n inputs and m outputs.

(iv) Multiplexers

A digital multiplexer, also known as MUX, is a logical circuit having 2^n inputs, one output and n selection variables or lines. The selection lines are used to make the output equal to the selected input.

(v) Demultiplexer

The converse of process of multiplexing is performed by demultiplexers, also known as DEMUX. In a DEMUX, there is one input and 2^n outputs and n selection lines which select the one output to be equal to the input.

(vi) Read-only memory (ROM)

A ROM is a device with n-to-2^n line decoder and m OR gates, each having 2^n inputs, all integrated in a single package. Such a device thus has n inputs and m outputs and is called a $2^n \times m$ ROM. The inputs to the OR gates are provided with links, that can be broken depending upon the desired outputs. Through these links all or some minterms can be applied to the OR gates.

The selection of OR gate links for desired outputs is called the programming of the ROM. The inputs numbering 2^n are called the addresses and the m outputs are called the output words. Thus for every address there is a corresponding word, which is available or read on the application of the address as input. Thus a ROM stores 2^n words each of m bits.

In a programmable ROM, all links are intact and they are removed by the programmer according to the output desired. Non-programmable ROMs are programmed during manufacturing. In Electrically Programmable ROMs, the programming can be altered by use of ultraviolet light.

(vi) Programmable logic arrays

A (n x p x m) PLA contains a group of p AND gates and m OR gates and has 2n inputs - made from n variables and their n complements. Each AND gate has all the 2n inputs through links and each OR gate has an input, the outputs of each of p AND gates. OR gate inputs are also through links. Thus n variables can provide m digit output, according to the programming of the PLA. Fig. 11.4 shows a (2 x 3 x 2) PLA.

11.3. SEQUENTIAL LOGIC DESIGNS

A sequential logic is that which provides output not only depending upon the inputs at that moment but also takes into consideration the state of the circuit at that moment. Thus the output is result of the effect of the inputs on the state, which depends upon the past history.

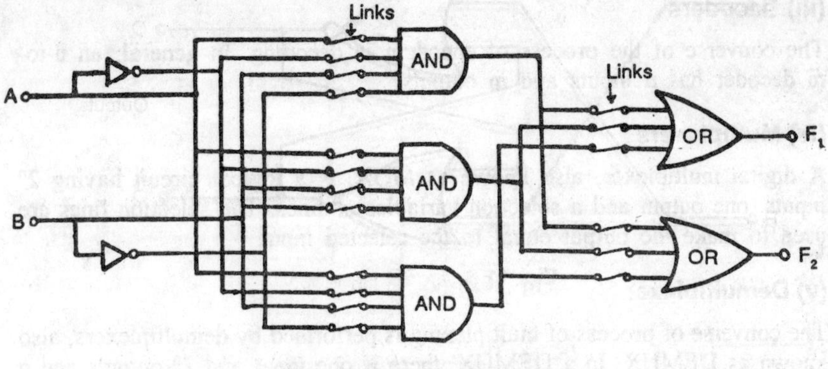

Fig. 11.4. (2 x 3 x 2) PLA.

A sequential logic essentially has a storage of memory elements to keep record of the state alongwith a combinational logic unit. Block diagram of a sequential logic is shown in Fig. 11.5.

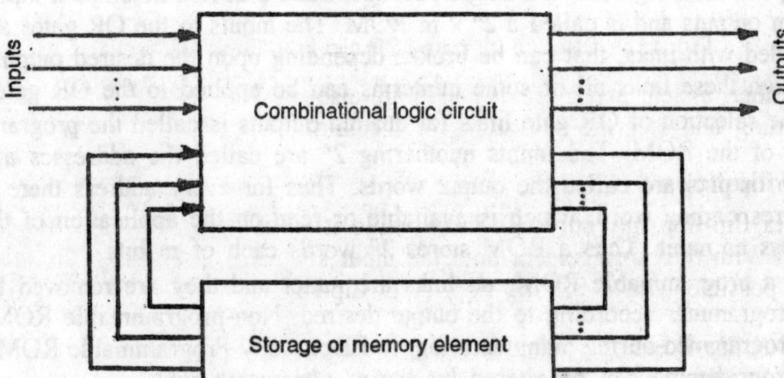

Fig. 11.5. General sequential logic circuit.

(i) Set-reset (SR) flip-flops

A bistable multivibrator is known as a flip-flop. A flip-flop stores 1 bit of information either a 0 or a 1 and provides two outputs - one equal to the stored bit and the other equal to its complement. One value of an output is called a set state and the other value of the same output is called the reset state. Thus for every change of input the circuit is set or reset alternately. Such a circuit is shown in Fig. 11.6. wherein two NOR gates have been used, with feedback to each other.

The above-described flip-flop circuit responds to every input, which may be a noise signal. To have a response at fixed time, clocked pulses could be applied to ensure output only when the clock pulse exists. Such an arrangement is shown in Fig. 11.7 alongwith its symbol.

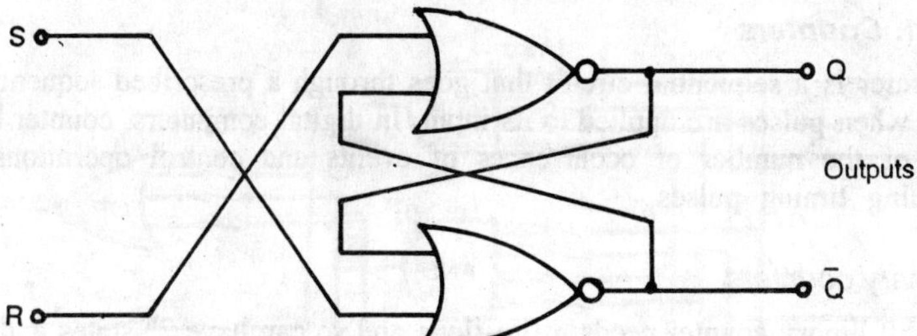

Fig. 11.6. An SR flip-slop.

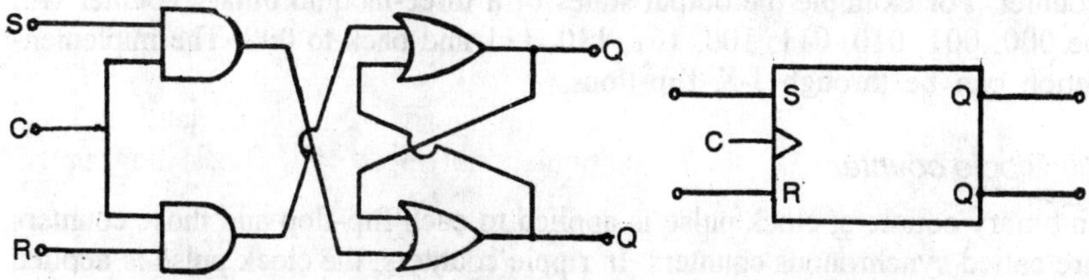

Fig. 11.7. Clocked (or gatel) SR clip-flap and its symbol.

Further, a clear and a preset pulses are also applied to the NOR gates for presetting and clearing operations of the unit.

(ii) D flip-flops

A data flip-flop has only one input while it complement makes the other input. With this arrangement both the input signals can never be the same as is possible with a SR flip-flop. A, D flip-flop is also clocked.

(iii) J-K flip-flops

A J-K flip-flop can have both the inputs equal. It is achieved by the use of two AND gates, each having one input and one feedback signal. Its circuit and the block diagram are shown in Fig. 11.8.

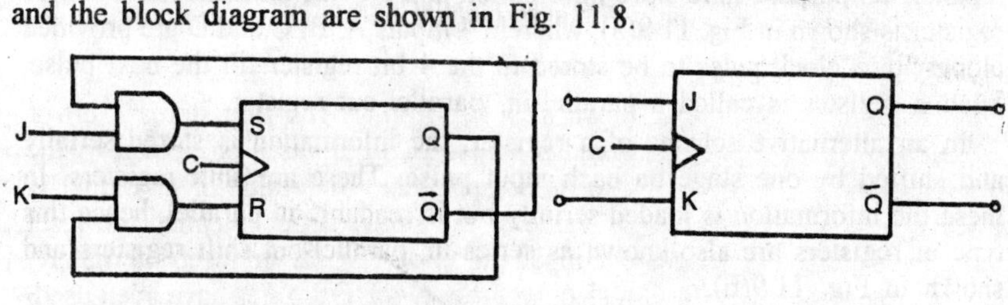

Fig. 11.8. J-K flip-flop.

11.4. Digital devices

Components used in digital computers such as counters, registers, memories and such other devices are described here briefly.

11.4.1. Counters

A counter is a sequential circuit that goes through a prescribed sequence of states when pulses are applied to its input. In digital computers, counter keep track of the number of occurrences of events and control operations by providing timing pulses.

(i) Binary counters

An n-bit binary counter needs n flip-flops and so can have 2^n states and can count from 0 to $(2^n - 1)$. The number of states m, of a binary counter, which is less than or equal to the possible states 2^n, is called the modulus of the counter. For example the output states of a three-modulo binary counter will be 000, 001, 010, 011, 100, 101, 110, 111 and back to 000. The implementation can be through J-K flip-flops.

(ii) Ripple counter

In binary counters, clock pulse is applied to each flip-flop and those counters are called synchronous counters. In ripple counters, the clock pulse is applied to only the first flip-flop while the output pulse of one flip-flop works as the triggering pulse for the next flip-flop. These are simpler than the binary counters but are slower and have the problem of glitches due propagation delays in the flip-flops.

(iii) Ring counters

These are synchronous counters and output of one flip-flop becomes the input of the next flip-flop making output of the last flip-flop the input for the first. Thus the pulses move in a cyclic order.

11.4.2. Registers

A register is a collection of flip-flops, equal in number to the bits of the register. Its purpose is to store information and provide on demand. A simple register is shown in Fig. 11.9(a), wherein 4 inputs A, B, C and D are provided alongwith a clock pulse to be stored in the 4 bit register till the next pulse. Such a register is called a parallel-in, parallel-out register.

In an alternative scheme of a register, the information is stored serially and shifted by one stage on each input pulse. These are shift registers. In these the information is loaded serially but is readable in parallel, hence this type of registers are also known as series-in, parallel-out shift registers and shown in Fig. 11.9(b).

11.4.3. Memories

In digital computers, memories store information in binary form. These may be composed of flip-flops, magnetic cores or capacitors.

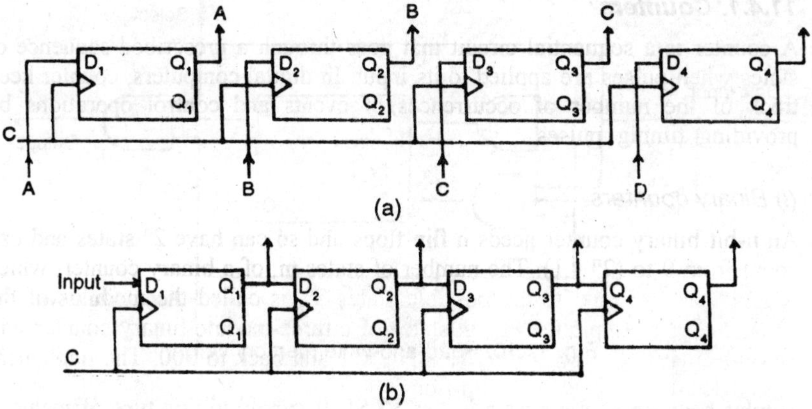

Fig. 11.9. Registers. (a) Parallel-in, parallel-out register. (b) Series-in, parallel-out register.

(i) Read-only memory (ROM)

We have already understood that a ROM has inputs as addresses and outputs as words. A ROM can be formed by using an n-to-2^n-line decoder whose output lines, known as word lines are connected to a matrix array of diodes connected in between the word lines and the output lines. As an address is fed as an input, its corresponding word appears on the word line, forward biasing the diode and hence providing the output.

ROMs with diodes may have a weak output. If the diodes are replaced by BIJs and made to conduct on a word signal from the decoder, the fixed bias voltage of the diode will appear as the output. For programming such a ROM, link to the BJT gate may be broken on selected locations.

Use of MOSFET in place of the diodes or BJIs, provides output on very weak inputs and the ROM becomes very compact, which is essential for large capacities.

(ii) Read-and-write memories (RAMs)

The memory which is known as the RAM in a computer system is in fact a read-and write-memory. Its one cell is shown in Fig. 11.10. It has three inputs - data, read/write and selection and one output. The cell can be written or read only when the select signal is on. In case of a read signal, the cell data is available at the output, while on a write signal the cell address stores data. Such memories are fabricated with either BJTs, NMOS or CMOS. Such memories are also known as static memories as the state of the flip-flop remains static till the next input pulse.

Another category o read/write memory is fabricated from NMOS a capacitor. The information signal is stored in the capacitor as its charge. Since the capacitor loses charge due to leakage, it requires a refreshing signal from time to time. Such a memory is called a dynamic RAM and is very

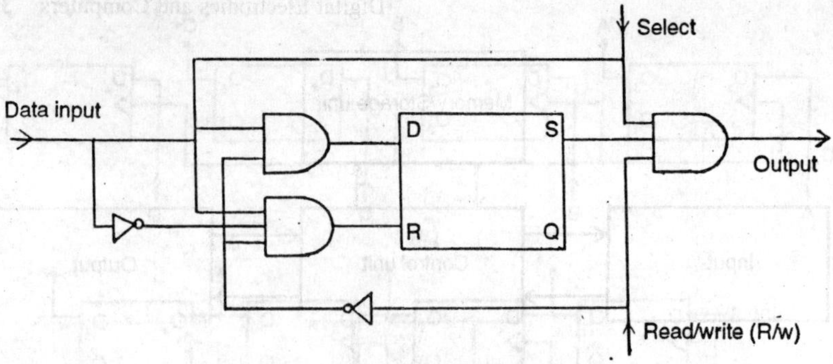

Fig. 11.10. Read-and-write memory cell.

popular because of the possibility of VLSI. If supply to this type of memory is cut off it loses information, while the discontinuance of supply to a static RAM does not change the state.

Another type of memory of this class is the magnetic memory in which small rings of ferrites are magnetised by flow of current through the conductor passing through them. Its readout is of destructive nature i.e. the information is lost on a readout.

(iii) Sequential memories

Magnetic floppy disks and tapes are the sequential memories. The information is permanently stored in such memories by magnetizing a cell on the surface of the device. The read/write operation is carried out by a magnetic head, whose coil gets a voltage by induction when the magnetization of its core is altered for a read operation. The reverse process occurs for write operation.

(iv) Charge-coupled memories

It is an enhancement MOSFET with numerous gates. The application of a positive voltage to the gate produces a potential well in the substrate. By applying a suitable voltages to different gates at different times, the potential wells may be shifted from below one gate to the other and thus stored. This is also a sertial memory as the stored information is available in sequence only.

11.5. DIGITAL COMPUTER

The block diagram of a digital computers working is shown in Fig. 11.11, which generally depicts how a digital computer performs its function of sequentially executing a set of instructions on data in digital form. Data for a computer, may be fed from the keyboard or mouse or stored in its memories. But the computer takes this data first to the RAM to process it. RAM is the main working field of the computer.

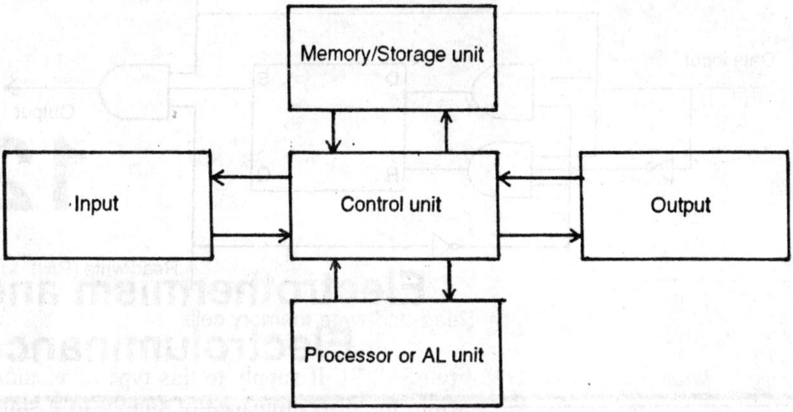

Fig. 11.11. Simplified digital computer.

The job of the processor or the arithmetic and logic unit (ALU) is to perform all arithmetic and logical operations, through various logic circuits. Intermediate results of logical operations are stored in ALU registers known as accumulators, which receive data from the RAM and sore it back in the RAM.

The control unit is the heart of the computer. After it reads and interprets an instruction, it sends the appropriate timing and control signals to that portion of the computer, which is supposed to perform that action. The control units accepts the instructions one by one is sequence and executes them with the help of counters, combinational and sequential logic circuits and temporary storage registers. The combination of control unit and the ALU is called the Central Processing Unit (CPU). A CPU on a single chip, as is the usual practice, is called the microprocessor.

12

Electrothermism and Electroluminance

Q.1. Copper can not be used as a heating element material.

(a) True (b) False

Q.2. For a heater operated at constant voltage, heat produced is proportional to :

(a) square of current,
(b) Resistance,
(c) Inverse of square of resistance,
(d) none of the above.

Q.3. Heating element materials, usually chosen, are of high resistivity for :

(a) more power to the heaters,
(b) economy of material,
(c) easier availability,
(d) none of the above.

Q.4. The main difference between the materials of resistors and heating elements is of :

(a) cost, (b) melting point,
(c) resistivity, (d) none of the above.

Q.5. When two electrodes in close proximity are immersed in water, the current passed between the electrodes, is -

(a) conduction current, (b) ionic current,
(c) convection current, (d) none of the above.

Q.6. Ice can be heated up by -

(a) induction heating, (b) dielectric heating,
(c) conduction heating, (d) none of the above.

Q.7. Silicon carbide heaters are popular for -
(a) high voltage applications,
(b) high current applications,
(c) high temperature applications,
(d) none of the above.

Q.8. In incandescent lamps, tungston is used for filament, mainly because of its :
(a) non-oxidising property, (b) high melting point,
(c) high resistivity, (d) none of the above.

Q.9. Electric current through metals produces heat because of -
(a) electronic vibrations, (b) dipole reorientations,
(c) ionic vibrations, (d) none of the above.

Q.10. A heating element design is fixed by -
(a) resistance, (b) resistance to diameter ratio,
(c) resistivity, (d) none of the above.

Q.11. A heating element for higher resistance to diameter ratio, must have -
(a) higher temperature coefficient to resistance,
(b) higher melting point,
(c) higher mechanical strength,
(d) none of the above.

Q.12. In dielectric heating, higher the frequency, lower the depth of heat penetration.
(a) True, (b) False.

Q.13. Induction heating can be used for heating :
(a) all metals, (b) only magnetic materials,
(c) all non-magnetic materials, (d) none of the above.

Q.14. Dielectric-heating can be used for heating
(a) all dielectrics,
(b) only monmagnetic materials,
(c) all higher resistivity materials,
(d) none of the above.

Q.15. Dielectric heating is possible with dc supply also.
(a) true, (b) false.

Q.16. A dielectric material gets heated when a dc voltage is applied across it through two plates on two sides. This type of heating may be categorized as -
(a) dielectric heating, (b) induction heating,
(c) resistive heating, (d) none of the above.

Q.17. Dielectric heating power is proportional to -
(a) frequency, (b) square of frequency,
(c) square-root of frequency, (d) none of the above.

Q.18. In ac fields, dielectric constant of a dielectric material is a -
(a) vector quantity, (b) scaler quantity,
(c) complex number, (d) none of the above.

Q.19. If a layer of air is also included between the dielectric heating plates alongwith a solid dielectric material layer, the air will get heated -
(a) more than the solid, (b) less than the solid,
(c) equal to the solid, (d) none of the above.

Q.20. Dielectric heating can be used to heat -
(a) only solid dielectrics,
(b) only solid and liquid dielectrics,
(c) solid, liquid and gaseous dielectrics.

Q.21. Depth of penetration of induction heat in magnetic metals, as compared to nonmagnetic metals will be -
(a) less, (b) more,
(c) same.

Q.22. Induction heating uses the principle of electromagnetic induction, as applicable to -
(a) transformers, (b) ac alternators,
(c) dc generators, (d) none of the above.

Q.23. Source of power for induction heating can be (i) ac supply mains, (ii) solid-state invertors, (iii) vacuum tube oscillators -
(a) all the above are true, (b) only (i) is false,
(c) only (ii) is false, (d) only (iii) is false.

Q.24. Welding process if basically a process of -
(a) resistive heating, (b) induction heating,
(c) dielectric heating, (d) none of the above.

Q.25. Welding process is applicable to -
(a) metals only, (b) conductors only,
(c) solids only, (d) none of the above.

Q.26. Resistance welding process is applicable to -
(a) conductors only,
(b) both conductors and insulators
(c) magnetic materials only
(d) none of the above

Q.27. Arc welding current is chosen depending upon -
(a) the electrode size, (b) metal thickness,
(c) source voltage, (d) none of the above.

Q.28. Arc welding source voltage has to be -
(a) ac only, (b) dc only,
(c) ac and dc both, (d) ac or dc, any.

Q.29. Copper wires are joined by -
(a) arc welding, (b) spark welding,
(c) resistance welding, (d) none of the above.

Q.30. Voltage of arc welding source, when open-circuited is of the order of -
(a) 400-500 V, (b) 100 to 300 V,
(c) 30-70 V, (d) none of the above.

Q.31. Ultraviolet heating radiations are of frequency -
(a) higher than light, (b) lower than light,
(c) same as light, (d) no relation with light.

Q.32. Infrared heating radiation wavelengths are -
(a) higher than light, (b) lower than light,
(c) same as light, (d) no relation with light.

Q.33. Velcoity of infrared radiations is -
(a) lower than light, (b) higher than light,
(c) same as light, (d) no relation with light.

Q.34. Infrared and ultra violet radiations are -
(a) electromagnetic waves, (b) transverse electric waves,
(c) transverse magnetic waves, (d) none of the above.

Q.35. Lesers beam frequency is -
(a) more than the visible spectrum,
(b) less than the visible spectrum,
(c) same as the visible spectrum.

Q.36. Basic properties of a photon are (i) direction of oscillation, (ii) phase of oscillation, (iii) frequency of oscillation.
(a) only (i) is false, (b) only (ii) is false,
(c) only (iii) is false, (d) all are true.

Q.37. Coherence of wave spectrum means equality of -
(i) direction of oscillations, (ii) phase of oscillation,
(iii) frequency of oscillation.
(a) only (i) is false, (b) only (ii) is false,
(c) only (iii) is false, (d) all are true.

Q.38. As compared to photons of ordinary light, photons of laser beams have -
(a) higher frequency range, (b) longer wavelengths,
(c) higher velocity, (d) none of the above.

Q.39. In contrast with ordinary light, laser light is characterised by -

(a) colour, (b) coherence,
(c) dispersion, (d) refraction.

Q.40. The coherence of laser beams is obtained by -

(a) stimulated emission, (b) wave guide,
(c) source of radiation, (d) none of the above.

Q.41. The reflective mirrors of a laser cavity -

(a) aid in coherence,
(b) aid in stimulation of emissions,
(c) dampen beam intensity,
(d) none of the above.

Q.42. Laser beams can be focused on a tiny spot because of their -

(a) high velocity, (b) low refractive index,
(c) high coherence, (d) none of the above.

Q.43. All types of electromagnetic waves have some coherence -

(a) true, (b) false.

Q.44. As compared to the coherence of ordinary light, coherence of lasers is of the order of -

(a) 3 times, (b) 7 times,
(c) 15 times, (d) none of the above.

Q.45. Laser beams obtained from different sources have -

(a) same properties, (b) different properties.

Q.46. Illumination spread around a source, in a specified and limited space is -

(a) uniform,
(b) varies inversely-proportional to the distance from the same,
(c) varies in proportion of the reflectivity of the object,
(d) none of the above.

Q.47. Discharge lamps are more efficient than the incandescent lamps due to -

(a) low heating effects, (b) photo-multiplier effect,
(c) none of the above.

Q.48. Ultraviolet are infrared rays have the same -

(a) velocity, (b) wavelength,
(c) frequency, (d) none of the above.

Q.49. Laser waves and the ordinary light lie in the spectrum of same (i) velocity, (ii) frequency, (iii) wavelength -

(a) only (i) is true, (b) only (i) is false,
(c) only (ii) is true, (d) all are true.

Q.50. Interference-patterns of light beams are obtained due to the pressure of -

(a) phase difference,
(b) coherence,
(c) phase difference and coherence,
(d) none of the above.

Q.51. Laser beams have (i) reflective, (ii) refractive properties -

(a) only (i) is true, (b) only (ii) is true,
(c) both are true, (d) none is true.

ABSTRACT

Electricity's is thermic effects are well known. Various commercial applications and devices use these thermic effects. The major heating processes by electricity are listed below, which uses different frequencies of power supply and are available in different capacities.

	Technique	*Frequency range*	*Power range*
(i)	Resistive	0-50 Hz	10 W-30 MW
(ii)	Induction heating	50 Hz-13.6 MHz	1 KW-10 MW
(iii)	Dielectric heating	1-100 MHz	1 KW-500 KW
(iv)	Lasers	30 THz	0.1-60 KW
(v)	Microwave heating	0.5-25 GHz	1-100 KW
(vi)	Arc melting	50 Hz	1-100 MW
(vii)	Electric welding	50 Hz	10-100 KW
(viii)	Infrared heating	1-1.5 THz	1-20 KW
(ix)	Ultraviolet heating	0.5-0.75 THz	1 KW
(x)	Illumination	50 Hz	10 W-1000 W

12.1. RESISTIVE HEATING

Heat developed in a resistive heating elements is given by -

$$P = I^2R \text{ W}$$

$$= V^2/R \text{ W}$$

Most of the supply systems being of constant voltage type, heat developed is inversely proportional to the resistance. The materials used for heating elements should have the following properties :

1. High melting point and stability at high temperatures.
2. No oxidation at high temperatures.
3. High resistivity for lower volume of resistor material for a given power rating.
4. Low temperature coefficient of resistance.

Two elemental heating materials are platinum and tungston, but these materials, being costly, are used in very special applications such as in electron tubes catode heators. In most other applications alloys or sintered products are used as heating elements. Some of these are listed below.

Material	*q*	*p*	*a*	*Principal applications*
Copper	40	1.73	-	Low-power surface heaters
Nickel-based alloys* :				
80Ni/20Cr	1200	108	6.0	Furnaces; resistance heaters
80Ni/20Cr + Al	1250	124	-2.0	Lower-cost furnaces
60Ni/15Cr/25Fe	1100	112	13	Lower-cost furnaces
50Ni/18Cr/32Fe	1075	111	17	Lower-cost furnaces
37Ni/18Cr/32Fe	1075	105	24	Lower-cost furnaces
44Ni/56Cu	400	49	-	Resistance haters; domestic appliances
Iron-based alloys*				
72Fe/22Cr/4Al	1050	139	4.7	Furnaces
72Fe/22Cr/4Al + Co	1375	145	3.2	Furnaces
78FFe/18Cr/4Al + Yt, C	1300	134	12	Furnaces
Refractory materials :				
platinum	1300	11	392	Small muffle furnaces
90Pt/10Rh	1550	19.2	200	Small muffle furnaces
60Pt/40Rh	1800	17.4	200	Small muffle furnaces
molybdenum	1750*	5.7	550	Vacuum furnaces : small muffle furnaces in hydrogen atmosphere
tantalum	2500	12.5	320	Yacuum furnaces
tungsten	1800*	5.6	594	Infrared lamps; vacuum furnaces
Non-metals :				
graphite	3000*	1000	-26.6	Vacuum furnaces and reducing atmospheres
molybdenum disilicide	1800	40	1200	Small glass-melting furnaces; forehearths
silicon carbon	1600	110000	-0.26	Furnaces; oxidising and reducing atmospheres
lanthanum chromite	1800	2100	-2.77	Furnaces; oxidising and reducing atmospheres

* Composition approximate. ** Not in air *** Maximum operating temperature in dry air (°C); p resistivity at 20°C (mWm × 10^{-2}); a mean resistance/temperature coefficient over operating range (x 10^{-5}).

12.2. INDUCTION HEATING

When a metal to be heated is a subject of electromagnetic induction for induced voltage and hence the flow of current, the process is said to be induction heating. From melting of iron and steel scrap in specially designed induction furnaces to sophisticated selective surface hardening of engineering components, steel tube continuous seam welding to heating a continuously moving steel rod for forging operations, induction heating is in a developed stage.

The induction heating, selection, generation and application of frequency is very critical as skin effect and proximity effect play important roles in the process of induction heating. Right from mains frequency for steel melting to 13.6 MHz for surface treatment is the range utilized for induction heating. The most important criteria are the desired heat-affected zone (HAZ) and the depth of penetration of heat in the object, for the selection of the frequency. Some typical applications and the frequencies are listed below:

1. Iron-core steel melting furnace	50-100 Hz
2. Coreless steel melting furnace	500-1000 Hz
3. Forging furnaces	1-10 KHz
4. Steel tube seam welding	400-500 KHz
5. Soldering, laboratory testing	1-50 MHz
6. Case hardening/surface heat treatment	5-13.6 MHz

For generation of frequencies upto 1000 Hz, rotary frequency converters, for 1000 to 100 KHz solid-state inverters and the higher frequencies vacuum tube oscillators are used.

12.3. DIELECTRIC HEATING

We have studied that when an alternating voltage is applied across a capacitor, the dielectric material is between the capacitor plates absorbs energy from the source. This is caused by the imaginary part of the dielectric constant of the material. This principle is utilized in dielectric heating for heating of dielectric materials. The must favourable feature of this process is the uniform heating of the material from surface to the core as the heat is developed in the material itself, at each spot. Manufacturing of plywood boards uses dielectric heating.

The frequencies applied in dielectric heating are of radio frequency range. The r.f. field produces sympathetic changes in the orientation of dipoles in the material. There changes in orientation give rise to frictional losses producing heat in the material. For uniform heating. The power dissipated in the material by this process is given by :

$$P = 2\pi f E^2 \varepsilon \tan \delta$$

where f = supply frequency

E = electric field strength

ε = Permitivity of the material

δ = Dissipation power angle

Hence the heat produced depends upon the characteristic properties of the material. Some materials which can be heated by this process are water, porcelain, glass, nylon, PVC, rubber, wood, paper, leather etc.

12.4. LASERS

The word LASER stands for 'light amplification of stimulated emission of radiation'. Normal light is the electromagnetic wave of wavelength 0.01 mm to 100 mm and the visible part of it has the wavelength of 0.4 mm to 0.75 mm, the two side bands are of ultraviolet and infrared light.

Light is composed of photons, each photon is an independent random state, while some of the lot many here the same state. The property of the photons to occupy similar state is known as coherence. The coherence is measured in time. Longer the time, more the coherence. Thus the coherence time reflects the period for which some photons are found in similar state. Ideally, for a polarized, monochromatic light this time should be infinite.

Laser beam has nearly 7 times the coherence of ordinary light. Its value is of the order of 10^{-3} s which means that the laser beam remain coherent during its travel of 100 km.

Photons is a quantum of energy which can be absorbed by some atoms to reach a higher level of energy, similarly, atom at higher level of energy may emit a photon to come back to lower energy level. This is called lasing transition.

For lasing action, the active media chosen is such that it has a majority atoms at higher energy level and when stimulated minority low-energy atoms absorbs energy and become high-energy atoms while the majority high-energy atoms undergo lasing transitions. These two transitions go on continuously with lasing transitions always more than the other transitions.

Optical resonant cavity

The photon emitted by the stimulated transition is in phase with the stimulating photon. Hence for the lasing transitions to emit photons of similar states, it is essential that the stimulating photons are limited to a particular state. The selection of some photon states and the suppression of the other states is realized by optical resonator, a principal component of each laser.

The general version of an official resonator, has a cylindrical hollow tube with mirrors on its two flat ends. One of the mirror is made semitransport. The active media is placed in between these two mirrors inside the tube. If it is solid, it is taken in a cylindrical shape with length nearly 10 times its diameter, is placed inside the cavity with its axis coinciding that of the cavity. This axis is the axis of the lasers.

An spontaneous emission of photons from the active material takes place Photons which have a direction parallel to the axis of the cavity keep on travelling between the two mirrors, with some part coming out as laser beam. The photons of other directions die out in the cavity. Thus photons of one direction only are selected, which keep in multiplying due to stimulated emissions.

Characteristics of optical cavity

(i) The quality factor

Q-factor is a dimensionless quantity used to show the ability of the cavity to store energy and is given by :

$$Q = 2\pi \omega t_c$$

where w is the central frequency of the lasers

t_c is the time constant of decay of the field energy in the resonant cavity. Slower the decay, higher the Q factor.

(ii) Losses

Irradiance suffers loses during travel inside the cavity and from the side walls, the instantaneous irradiance given by :

$$I(z) = I_o e^{-az}$$

where Z is the distance of travel in Z direction

I_o is the irradiance at origin

a is the loss factor in per meter

$$Q = \frac{2\pi}{\lambda a}$$

where λ is the wavelength.

(iii) Radiant losses

The laser energy leaving the semitransparent aperture of the cavity is also a loss to the cavity energy. This loss factors is given :

$$a_{rad} = \frac{1}{2LR}$$

where L is the resonator length

R the reflectivity of the output mirror

$$= \frac{\text{Reflected light}}{\text{Total incident light}}$$

(iv) Diffraction losses

The light energy suffers loss due to diffraction at the edges of the mirrors. It is given by Fresnel number :

$$N_F = \frac{a^2}{4 \pi L}$$

where a is the mirror diameter.

Leser bean characteristics

(i) Gain coefficient

As the light flux travels in the active material, it stimulates emissions and the flux keeps on multiplying during its travel. The increment of light flux is proportional to the initial irradiance in the z direction of beam travel and is given by :

$$\Delta I = \beta I(z) \, \Delta z$$

where β is called the gain coefficient.

And the power generated per unit volume of the active material is given by $P(z) = \beta \, I(z)$

Also β is given by $\beta = \sigma_{12} \, (n_2 - n_1)$

where σ_{12} = stimulated emission cross-section

$$= \frac{\text{emission process probability}}{\text{irradiance of the flux of photons}}$$

$(n_2 - n_1)$ = population difference of higher level energy atoms and lower energy level atoms.

Laser application

Lasers find applications in scientific research and development, manufacturing of precision electronic components, clinical operations, navigation and material processing.

Due to its high concentration of energy at a tiny spot, it is an ideal tool for precision working such as semiconductor etching, neural surgery and drilling a hole in metal of micrometer order.

In material processing, cutting, melting, heat treatment, surface coating, welding, soldering, brazing etc. are large number of applications of lasers.

12.5. MICROWAVE HEATING

Microwave ovens have now entered indian kitchens due to their instant heating operation without burnouts, uniform distribution of heat through the heated material without stirring it.

Microwave ovens make use of devices such as magnetron and klystron. The largest size of magnetron available is of 25 KW at 296 MHz with ratings reduced to 5KW at 2450 MHz. Various types of microwave applicators have been developed for industrial heating processes, the much effective type is of serpentine shape. The resonant energy is recirculated in the heating process.

The microwaves are not safe to humans hence adequate safety precautions are needed in this process.

Textile, rubber, heather, food and pharmaceutical industry employ microwave heating.

12.6. ARC MELTING

Electric are furnaces use electric arc produced between graphite electrode and the metal pool to melt the metals in steel industry. Three phase direct arc method is the most popular method adopted.

Usually ac supply of 1.5 to 3 volt of thousands of ampe[illegible] is used for production of arc in the furnace which may be upto a diameter of 7 meters with 400 MT of charge capacity. DC arc has the possibility of better utilization of furnace but the power supply is the problem.

The furnaces have provision of vertical movement of graphite electrodes for control of arc current and balance of three phases. Special transformers, having capacity to tolerate large mechanical stresses due to short circuits are used to supply power to the furnaces.

Indirect arc is also used to heat the metals where in an arc between two electrodes supply heat by radiation. Such furnace efficiency is low and have been manufactured upto a capacity of 1 MT of 1 MW power rating.

12.7. ELECTRIC WELDING

The largest and the most widely uses of electrothermics is done in electric welding. A large variety of welding processes have been developed and are in use depending upon the applications. These are :

A. Arc welding :
- (a) direct arc welding
- (b) Submerged arc welding
- (c) Electroslag welding
- (d) Tungston inert gas (TIG) welding
- (e) Plasma arc welding
- (f) Gas shielded metal arc welding

B. Resistance welding :
- (a) Spot welding
- (b) Projection welding
- (c) Seam welding
- (d) Butt and flash welding

In arc welding, usually an electric arc is striked by short circuiting an electrode with the object metal piece and then withdrawing the electrode a little to maintain the arc. The arc voltage is usually of the order of 2 volts. Such low voltage is supplied by a transformer to high leakag[illegible] reactance with drooping voltage current characteristic.

The resistance welding processes require the metals to be welded to be processed together under two electrodes having a low-voltage supply. The junction resistance of the metals produces intense heat and melts the junction point on the two sides, fusing them together. It is a fast process applicable to sheet metal working.

12.8. INFRARED HEATING

Infrared radiations caused by heating of heating elements can be focused on to an object material to heat it. Various shapes and types of infrared sources are :

(i) Ceramic heaters
(i) Mineral insulated heaters
(iii) Tubular quartz-sheathed heaters
(iv) Circular heat lamps
(v) Linear heat lamps
(vi) Quartz halogen heat lamps

As the temperature of the source is increased more infrared radiation takes place. The source temperatures range from 800°C to 2700°C giving radiation densities of 68 to 300 kw/m^2. Infrared heaters are used for heating and drying of non-metallic materials.

12.9. ULTRA-VIOLET HEATING

Inks and plastic can be cured at high rates with ultraviolet heating. The heat density of ultraviolet light is very low. By heating a source of heat all types of radiations are produced. For getting only ultraviolet light, all other radiations are absorbed in the envelop around the source.

Illumination

Illumination is basically a process in which light photons are emitted from a source, due to heating, electric discharge or florescence of a phosphor coating by ultraviolet energy.

Light is an electromagnetic radiation in the range of 380-400 nm and 760-780 nm. This range depends upon the capability upto observer, and hence not rigidly fixed.

Luminous flux is rate of flow of light. It varies with the direction of emission. The rate of change of luminous flux with solid angle is termed as luminous intensity.

Illumination is the process whereby luminous flux is incident upon a solid surface and the corresponding quantity (flux density per unit area) is the alluminance. Brightness of a luminous surface is termed as luminance.

Lamps

In lamps illuminescence is produced by heating the filament by flow of electric current through it.

Discharge lamps

A tube of glass, quartz or a suitable material, having a gas or a metal vapour is having two electrodes at the two ends. An electric current is passed through the tube creating a discharge in the gas producing light. Mercury

lamps, mercury blended lamps, mercury-halide lamps and sodium lamps are the examples of discharge tubes.

Properties of lighting

(i) Uniformity
(ii) Reflectance
(iii) Illuminance ratio
(iv) Modelling and shadow effects
(v) Glare, and
(vi) Colour rendering

Estimation of lighting

$$\text{Lumens required} = \frac{(\text{lm/m}^2) \times \text{area (m}^2) \times \text{waste light factor}}{\text{beam factor x maintenance factor}}$$

with recommended values as :

Office lighting	450 lm/m^2
Flood lighting	200-1000 lm/m^2
Shop lighting	500-1000 lm/m^2
Domestic general	200-400 lm/m^2
Specific	400-500 lm/m^2

Answers to Questionaires

1. Basic Principles

1. (b)	2. (c)	3. (b)	4. (a)	5. (c)
6. (c)	7. (a)	8. (a)	9. (b)	10. (b)
11. (d)	12. (c)	13. (a)	14. (d)	15. (c)
16. (a)	17. (d)	18. (d)	19. (c)	20. (c)
21. (c)	22. (b)	23. (b)	24. (a)	25. (b)
26. (b)	27. (b)	28. (c)	29. (c)	30. (c)
31. (d)	32. (b)	33. (c)	34. (a)	35. (d)
36. (c)	37. (a)	38. (c)	39. (d)	40. (b)
41. (d)	42. (d)	43. (d)	44. (b)	45. (a)
46. (a)	47. (d)	48. (d)	49. (a)	50. (b)
51. (d)	52. (a)	53. (c)	54. (a)	55. (a)
56. (b)	57. (b)	58. (d)	59. (a)	60. (a)
61. (a)	62. (a)	63. (a)	64. (b)	65. (a)
66. (b)	67. (d)	68. (d)	69. (b)	70. (a)
71. (a)	72. (d)	73. (b)	74. (b)	75. (a)
76. (c)	77. (c)	78. (c)	79. (b)	80. (a)
81. (a)	82. (b)	83. (b)	84. (b)	85. (b)
86. (c)	87. (a)	88. (b)	89. (c)	90. (a)
91. (b)	92. (b)	93. (a)	94. (c)	95. (c)
96. (b)	97. (b)	98. (b)	99. (a)	100. (b)
101. (a)	102. (b)	103. (c)	104. (c)	105. (c)
106. (c)	107. (c)	108. (c)	109. (b)	110. (b)
111. (b)	112. (a)	113. (a)	114. (d)	115. (a)
116. (b)	117. (c)	118. (a)	119. (b)	120. (b)
121. (c)	122. (d)	123. (a)	124. (b)	125. (b)
126. (c)	127. (d)	128. (c)	129. (d)	130. (b)
131. (b)	132. (a)	133. (b)	134. (b)	135. (a)

136. (b)	137. (b)	138. (b)	139. (c)	140. (c)
141. (b)	142. (c)	143. (b)	144. (c)	145. (b)
146. (a)	147. (c)	148. (d)	149. (d)	150. (a)
151. (a)				

2. Materials and Properties

1. (b)	2. (d)	3. (b)	4. (b)	5. (b)
6. (a)	7. (b)	8. (c)	9. (a)	10. (b)
11. (b)	12. (c)	13. (b)	14. (b)	15. (c)
16. (d)	17. (c)	18. (a)	19. (a)	20. (c)
21. (d)	22. (b)	23. (c)	24. (a)	25. (b)
26. (b)	27. (b)	28. (b)	29. (b)	30. (b)
31. (a)	32. (b)	33. (a)	34. (b)	35. (a)
36. (a)	37. (b)	38. (a)	39. (a)	40. (a)
41. (a)	42. (a)	43. (a)	44. (b)	45. (b)
46. (b)	47. (b)	48. (c)	49. (a)	50. (b)
51. (b)	52. (c)	53. (a)	54. (b)	55. (d)
56. (d)	57. (c)	58. (c)	59. (a)	60. (b)
61. (c)	62. (b)	63. (b)	64. (a)	65. (b)
66. (a)	67. (b)	68. (a)	69. (a)	70. (b)
71. (c)	72. (b)	73. (c)	74. (a)	75. (a)
76. (b)	77. (d)	78. (c)	79. (b)	80. (a)
81. (c)	82. (a)	83. (b)	84. (c)	85. (b)
86. (b)	87. (c)	88. (c)	89. (a)	90. (b)
91. (b)	92. (c)	93. (b)	94. (a)	95. (a)
96. (b)	97. (b)	98. (b)	99. (b)	100. (c)
101. (b)	102. (b)	103. (d)	104. (a)	105. (c)
106. (b)	107. (c)	108. (d)	109. (a)	110. (b)
111. (c)	112. (a)	113. (b)	114. (a)	

3. Basic Principles

1. (b)	2. (a)	3. (c)	4. (b)	5. (b)
6. (a)	7. (c)	8. (b)	9. (c)	10. (a)
11. (a)	12. (b)	13. (c)	14. (b)	15. (b)
16. (c)	17. (b)	18. (a)	19. (b)	20. (a)
21. (b)	22. (a)	23. (a)	24. (b)	25. (a)
26. (b)	27. (c)	28. (b)	29. (b)	30. (b)

31. (b)	32. (b)	33. (c)	34. (b)	35. (a)
36. (a)	37. (c)	38. (b)	39. (b)	40. (d)
41. (a)	42. (a)	43. (c)	44. (a)	45. (a)
46. (b)	47. (b)	48. (a)	49. (a)	50. (c)
51. (b)	52. (b)	53. (b)	54. (a)	55. (a)
56. (c)	57. (c)	58. (c)	59. (d)	60. (b)
61. (b)	62. (c)	63. (b)	64. (a)	65. (c)
66. (b)	67. (a)	68. (a)	69. (b)	70. (a)
71. (d)	72. (b)	73. (a)	74. (c)	75. (a)
76. (c)	77. (a)	78. (a)	79. (a)	80. (b)
81. (c)	82. (d)	83. (a)	84. (b)	85. (b)
86. (d)	87. (a)	88. (d)	89. (d)	90. (a)
91. (a)	92. (c)	93. (d)	94. (b)	95. (b)
96. (b)	97. (a)	98. (c)	99. (a)	100. (b)
101. (b)	102. (d)	103. (c)	104. (b)	105. (b)
106. (b)	107. (c)	108. (b)	109. (c)	110. (b)

4. Control Engineering

1. (c)	2. (c)	3. (d)	4. (c)	5. (c)
6. (c)	7. (d)	8. (c)	9. (b)	10. (a)
11. (a)	12. (a)	13. (b)	14. (b)	15. (c)
16. (a)	17. (a)	18. (d)	19. (b)	20. (a)
21. (a)	22. (b)	23. (a)	24. (b)	25. (a)
26. (a)	27. (a)	28. (b)	29. (b)	30. (a)
31. (b)	32. (a)	33. (a)	34. (a)	35. (a)
36. (a)	37. (a)	38. (b)	39. (a)	40. (a)
41. (b)	42. (c)	43. (b)	44. (c)	45. (a)
46. (b)	47. (b)	48. (d)	49. (b)	50. (b)
51. (a)	52. (b)	53. (c)	54. (a)	55. (b)
56. (b)	57. (d)	58. (d)	59. (b)	60. (c)
61. (c)	62. (a)	63. (b)	64. (c)	65. (b)
66. (b)	67. (d)	68. (b)	69. (b)	70. (b)
71. (c)	72. (b)	73. (b)	74. (a)	75. (a)
76. (a)	77. (b)	78. (a)	79. (b)	80. (a)
81. (b)	82. (a)	83. (c)	84. (b)	85. (a)
86. (c)	87. (a)	88. (b)	89. (a)	90. (c)

91. (b)	92. (a)	93. (b)	94. (a)	95. (c)
96. (b)	97. (b)	98. (b)	99. (c)	100. (b)
101. (a)	102. (b)	103. (d)	104. (b)	105. (a)
106. (c)	107. (a)	108. (a)	109. (b)	110. (b)
111. (c)	112. (a)	113. (a)	114. (a)	115. (b)
116. (c)	117. (b)	118. (a)	119. (c)	120. (c)
121. (a)	122. (a)	123. (b)	124. (a)	125. (b)
126. (b)	127. (b)	128. (c)	129. (a)	130. (a)
131. (b)	132. (c)	133. (b)	134. (a)	135. (a)
136. (b)	137. (b)	138. (a)	139. (a)	140. (a)
141. (a)	142. (d)	143. (a)	144. (a)	145. (b)
146. (b)	147. (b)	148. (c)	149. (a)	150. (c)
151. (b)	152. (b)	153. (b)	154. (a)	155. (a)
156. (b)	157. (b)	158. (a)	159. (c)	160. (b)
161. (d)	162. (a)	163. (b)	164. (c)	165. (a)
166. (b)	167. (b)	168. (a)	169. (b)	170. (a)
171. (a)	172. (a)	173. (b)	174. (c)	175. (b)
176. (a)	177. (a)	178. (b)	179. (b)	180. (a)
181. (b)	182. (a)	183. (c)	184. (c)	185. (b)
186. (b)	187. (c)	188. (a)	189. (b)	190. (b)
191. (b)	192. (a)	193. (b)	194. (b)	195. (a)
196. (a)	197. (a)	198. (a)	199. (a)	200. (a)
201. (c)				

5. Power Systems

1. (b)	2. (b)	3. (b)	4. (b)	5. (a)
6. (a)	7. (b)	8. (b)	9. (a)	10. (d)
11. (b)	12. (c)	13. (a)	14. (a)	15. (c)
16. (b)	17. (c)	18. (b)	19. (a)	20. (a)
21. (d)	22. (c)	23. (a)	24. (a)	25. (a)
26. (a)	27. (a)	28. (a)	29. (b)	30. (a)
31. (b)	32. (b)	33. (a)	34. (a)	35. (b)
36. (a)	37. (a)	38. (a)	39. (a)	40. (b)
41. (a)	42. (b)	43. (a)	44. (a)	45. (a)
46. (a)	47. (a)	48. (a)	49. (a)	50. (a)
51. (a)	52. (a)	53. (a)	54. (a)	55. (b)

56. (b) 57. (a) 58. (c) 59. (a) 60. (a)
61. (a) 62. (b) 63. (b) 64. (a) 65. (a)
66. (a) 67. (b) 68. (a) 69. (b) 70. (a)
71. (d) 72. (a) 73. (a) 74. (a) 75. (a)
76. (a) 77. (b) 78. (a) 79. (a) 80. (a)
81. (d) 82. (a) 83. (b) 84. (a) 85. (b)
86. (a) 87. (a) 88. (a) 89. (a) 90. (a)
91. (a) 92. (a) 93. (b) 94. (a) 95. (a)
96. (b) 97. (c) 98. (b) 99. (b) 100. (d)
101. (d) 102. (b) 103. (b) 104. (a) 105. (a)
106. (a) 107. (b) 108. (a) 109. (b) 110. (c)
111. (b) 112. (b) 113. (b) 114. (a) 115. (b)
116. (d) 117. (a) 118. (b) 119. (b) 120. (a)
121. (a) 122. (b) 123. (b) 124. (a) 125. (b)
126. (c) 127. (c) 128. (a) 129. (b) 130. (a)
131. (a) 132. (b) 133. (b) 134. (b) 135. (b)
136. (b) 137. (b) 138. (b) 139. (b) 140. (a)
141. (d) 142. (a) 143. (b) 144. (a) 145. (c)
146. (a) 147. (c) 148. (c) 149. (d) 150. (a)
151. (b) 152. (c) 153. (b) 154. (c) 155. (a)
156. (a) 157. (b) 158. (c) 159. (c) 160. (c)
161. (c) 162. (c) 163. (c) 164. (b) 165. (b)
166. (a) 167. (b) 168. (b) 169. (a) 170. (a)
171. (a) 172. (b) 173. (c) 174. (b) 175. (a)
176. (a) 177. (c) 178. (a) 179. (a) 180. (c)
181. (c) 182. (a) 183. (c) 184. (a) 185. (c)
186. (a) 187. (b) 188. (a) 189. (c) 190. (a)
191. (c) 192. (c) 193. (c) 194. (a) 195. (c)
196. (d) 197. (c) 198. (c) 199. (a) 200. (b)
201. (a) 202. (b) 203. (c) 204. (c) 205. (b)
206. (a) 207. (c) 208. (a) 209. (b) 210. (c)
211. (b) 212. (a) 213. (c) 214. (a) 215. (b)
216. (a) 217. (a) 218. (b) 219. (c) 220. (a)
221. (a) 222. (a) 223. (b) 224. (a) 225. (a)
226. (c) 227. (b)

5. Electric Machines

1. (d)	2. (a)	3. (b)	4. (c)	5. (d)
6. (c)	7. (c)	8. (c)	9. (c)	10. (a)
11. (b)	12. (b)	13. (b)	14. (a)	15. (a)
16. (a)	17. (c)	18. (a)	19. (b)	20. (d)
21. (d)	22. (d)	23. (a)	24. (b)	25. (a)
26. (a)	27. (b)	28. (a)	29. (c)	30. (b)
31. (d)	32. (c)	33. (b)	34. (d)	35. (d)
36. (c)	37. (c)	38. (a)	39. (b)	40. (c)
41. (b)	42. (d)	43. (c)	44. (a)	45. (d)
46. (b)	47. (c)	48. (a)	49. (c)	50. (c)
51. (c)	52. (d)	53. (a)	54. (d)	55. (c)
56. (d)	57. (c)	58. (c)	59. (c)	60. (b)
61. (c)	62. (d)	63. (b)	64. (b)	65. (d)
66. (a)	67. (b)	68. (a)	69. (a)	70. (b)
71. (b)	72. (c)	73. (d)	74. (a)	75. (b)
76. (a)	77. (a)	78. (d)	79. (b)	80. (a)
81. (d)	82. (c)	83. (b)	84. (c)	85. (c)
86. (a)	87. (d)	88. (d)	89. (a)	90. (b)
91. (a)	92. (a)	93. (a)	94. (a)	95. (a)
96. (a)	97. (d)	98. (a)	99. (c)	100. (a)
101. (a)	102. (a)	103. (a)	104. (a)	105. (a)
106. (a)	107. (a)	108. (a)	109. (a)	110. (b)
111. (c)	112. (b)	113. (b)	114. (b)	115. (b)
116. (b)	117. (b)	118. (b)	119. (a)	120. (b)
121. (c)	122. (c)	123. (b)	124. (b)	125. (a)
126. (b)	127. (a)	128. (b)	129. (c)	130. (a)
131. (a)	132. (a)	133. (b)	134. (c)	135. (b)
136. (a)	137. (a)	138. (a)	139. (b)	140. (b)
141. (a)	142. (d)	143. (a)	144. (a)	145. (c)
146. (b)	147. (a)	148. (a)	149. (b)	150. (b)
151. (b)	152. (b)	153. (b)	154. (b)	155. (c)
156. (a)	157. (b)	158. (b)	159. (b)	160. (c)
161. (b)	162. (c)	163. (b)	164. (b)	165. (b)
166. (b)	167. (b)	168. (a)	169. (a)	170. (b)

171. (a) 172. (c) 173. (b) 174. (c) 175. (b)
176. (c) 177. (b) 178. (a) 179. (b) 180. (b)
181. (b) 182. (a) 183. (b) 184. (b) 185. (c)
186. (d) 187. (b) 188. (a) 189. (a) 190. (a)
191. (b) 192. (a) 193. (a) 194. (c) 195. (c)
196. (b) 197. (a) 198. (b) 199. (a) 200. (b)
201. (d) 202. (a) 203. (b) 204. (d) 205. (a)
206. (a) 207. (a) 208. (a) 209. (a) 210. (b)
211. (a) 212. (c) 213. (b) 214. (d) 215. (c)
216. (d) 217. (b) 218. (b) 219. (c) 220. (c)
221. (d) 222. (d) 223. (c) 224. (c) 225. (b)
226. (c) 227. (a) 228. (c) 229. (a) 230. (b)
231. (b) 232. (b) 233. (c) 234. (b) 235. (b)
236. (d) 237. (a) 238. (c) 239. (d) 240. (d)
241. (b) 242. (b) 243. (a) 244. (b) 245. (a)
246. (a) 247. (a) 248. (d) 249. (b) 250. (b)
251. (b)

7. Electric Drives and Traction

1. (a) 2. (c) 3. (c) 4. (a) 5. (b)
6. (c) 7. (a) 8. (b) 9. (b) 10. (d)
11. (a) 12. (c) 13. (b) 14. (b) 15. (b)
16. (c) 17. (b) 18. (b) 19. (b) 20. (a)
21. (b) 22. (c) 23. (b) 24. (b) 25. (c)
26. (a) 27. (a) 28. (b) 29. (b) 30. (a)
31. (b) 32. (a) 33. (b) 34. (d) 35. (b)
36. (b) 37. (c) 38. (a) 39. (b) 40. (d)
41. (c) 42. (b) 43. (b) 44. (c) 45. (c)
46. (a) 47. (b) 48. (c) 49. (b) 50. (c)
51. (a) 52. (c) 53. (c) 54. (c) 55. (a)
56. (c) 57. (a) 58. (c) 59. (c) 60. (a)
61. (c) 62. (a) 63. (a) 64. (b) 65. (b)
66. (a) 67. (c) 68. (a) 69. (b) 70. (a)
71. (c) 72. (c) 73. (b) 74. (b) 75. (b)
76. (a) 77. (c) 78. (b) 79. (a) 80. (a)
81. (b) 82. (c) 83. (a) 84. (b) 85. (b)

86. (a)	87. (b)	88. (b)	89. (b)	90. (c)
91. (c)	92. (c)	93. (b)	94. (b)	95. (a)
96. (a)	97. (a)	98. (c)	99. (a)	100. (c)
101. (d)				

8. Measurements and Instrumentation

1. (c)	2. (c)	3. (a)	4. (b)	5. (d)
6. (b)	7. (a)	8. (c)	9. (a)	10. (b)
11. (a)	12. (a)	13. (a)	14. (c)	15. (b)
16. (a)	17. (a)	18. (a)	19. (b)	20. (a)
21. (b)	22. (b)	23. (d)	24. (c)	25. (b)
26. (a)	27. (a)	28. (a)	29. (c)	30. (b)
31. (b)	32. (c)	33. (b)	34. (a)	35. (c)
36. (b)	37. (b)	38. (b)	39. (a)	40. (a)
41. (b)	42. (a)	43. (b)	44. (b)	45. (b)
46. (c)	47. (d)	48. (b)	49. (b)	50. (b)
51. (a)	52. (b)	53. (c)	54. (a)	55. (c)
56. (c)	57. (a)	58. (c)	59. (c)	60. (c)
61. (c)	62. (a)	63. (a)	64. (b)	65. (b)
66. (b)	67. (a)	68. (b)	69. (c)	70. (b)
71. (c)	72. (c)	73. (c)	74. (a)	75. (c)
76. (d)	77. (c)	78. (a)	79. (b)	80. (a)
81. (a)	82. (c)	83. (a)	84. (b)	85. (a)
86. (a)	87. (a)	88. (b)	89. (c)	90. (b)
91. (a)	92. (a)	93. (a)	94. (b)	95. (b)
96. (b)	97. (b)	98. (a)	99. (c)	100. (c)
101. (c)	102. (b)	103. (b)	104. (a)	

9. Electromagnetic Waves and Telecommunication

1. (a)	2. (a)	3. (a)	4. (c)	5. (b)
6. (a)	7. (a)	8. (c)	9. (a)	10. (b)
11. (b)	12. (a)	13. (a)	14. (a)	15. (c)
16. (b)	17. (a)	18. (a)	19. (a)	20. (b)
21. (b)	22. (a)	23. (b)	24. (d)	25. (c)
26. (b)	27. (b)	28. (c)	29. (c)	30. (c)
31. (b)	32. (a)	33. (b)	34. (c)	35. (b)

36. (b)	37. (b)	38. (c)	39. (b)	40. (c)
41. (a)	42. (c)	43. (c)	44. (b)	45. (c)
46. (a)	47. (b)	48. (c)	49. (b)	50. (a)
51. (b)	52. (b)	53. (b)	54. (a)	55. (b)
56. (a)	57. (c)	58. (a)	59. (b)	60. (b)
61. (c)	62. (a)	63. (b)	64. (a)	65. (c)
66. (b)	67. (b)	68. (b)	69. (b)	70. (b)
71. (c)	72. (a)	73. (a)	74. (a)	75. (b)
76. (b)	77. (a)	78. (b)	79. (b)	80. (b)
81. (a)	82. (a)	83. (a)	84. (a)	85. (b)
86. (c)	87. (c)	88. (c)	89. (b)	90. (b)
91. (c)	92. (c)	93. (b)	94. (c)	95. (d)
96. (d)	97. (a)	98. (a)	99. (a)	100. (b)
101. (b)	102. (a)	103. (b)	104. (b)	105. (a)
106. (a)	107. (a)	108. (b)	109. (b)	110. (b)
111. (a)	112. (c)	113. (a)	114. (c)	115. (b)
116. (b)	117. (a)	118. (b)	119. (c)	120. (a)

10. Power Electronics

1. (b)	2. (b)	3. (b)	4. (c)	5. (b)
6. (b)	7. (c)	8. (a)	9. (a)	10. (d)
11. (a)	12. (a)	13. (c)	14. (c)	15. (b)
16. (b)	17. (c)	18. (a)	19. (b)	20. (a)
21. (a)	22. (a)	23. (a)	24. (a)	25. (b)
26. (b)	27. (b)	28. (c)	29. (a)	30. (a)
31. (b)	32. (b)	33. (b)	34. (c)	35. (b)
36. (a)	37. (b)	38. (b)	39. (b)	40. (b)
41. (b)	42. (a)	43. (a)	44. (b)	45. (b)
46. (a)	47. (b)	48. (a)	49. (b)	50. (c)
51. (c)	52. (b)	53. (d)	54. (a)	55. (b)
56. (a)	57. (b)	58. (c)	59. (c)	60. (c)
61. (b)	62. (c)	63. (a)	64. (b)	65. (c)
66. (c)	67. (c)	68. (b)	69. (b)	70. (c)
71. (a)	72. (b)	73. (c)	74. (a)	75. (a)
76. (d)	77. (c)	78. (b)	79. (c)	80. (b)
81. (b)	82. (a)	83. (b)	84. (a)	85. (b)

86. (c)	87. (a)	88. (b)	89. (b)	90. (c)
91. (a)	92. (d)	93. (a)	94. (a)	95. (b)
96. (a)	97. (a)	98. (b)	99. (a)	100. (a)
101. (a)				

11. Digital Electronics and Computers

1. (a)	2. (b)	3. (a)	4. (b)	5. (b)
6. (b)	7. (b)	8. (b)	9. (a)	10. (b)
11. (a)	12. (c)	13. (b)	14. (b)	15. (c)
16. (c)	17. (a)	18. (a)	19. (b)	20. (a)
21. (b)	22. (c)	23. (b)	24. (a)	25. (b)
26. (b)	27. (b)	28. (b)	29. (a)	30. (b)
31. (a)	32. (b)	33. (b)	34. (b)	35. (c)
36. (c)	37. (a)	38. (a)	39. (b)	40. (b)
41. (b)	42. (b)	43. (b)	44. (b)	45. (b)
46. (b)	47. (a)	48. (b)	49. (b)	50. (a)
51. (b)	52. (b)	53. (a)	54. (b)	55. (a)
56. (c)	57. (b)	58. (c)	59. (b)	60. (c)
61. (c)	62. (a)	63. (b)	64. (c)	65. (c)
66. (c)	67. (b)	68. (d)	69. (b)	70. (b)
71. (b)	72. (b)	73. (b)	74. (b)	75. (a)
76. (b)	77. (b)	78. (a)	79. (a)	80. (a)
81. (a)	82. (b)	83. (c)	84. (b)	85. (c)
86. (c)	87. (a)	88. (b)	89. (b)	90. (b)
91. (b)	92. (a)	93. (a)	94. (a)	95. (c)
96. (a)	97. (c)	98. (a)	99. (c)	100. (b)
101. (c)	102. (b)	103. (b)	104. (b)	105. (c)
106. (a)	107. (a)	108. (c)	109. (a)	110. (a)
111. (a)	112. (c)	113. (a)	114. (c)	115. (b)
116. (b)	117. (b)	118. (c)	119. (a)	120. (c)
121. (c)	122. (b)	123. (c)	124. (b)	125. (c)
126. (c)	127. (c)	128. (b)	129. (c)	130. (b)
131. (a)	132. (c)	133. (c)	134. (a)	135. (c)
136. (a)	137. (b)	138. (b)	139. (b)	140. (b)

141. (b)	142. (a)	143. (a)	144. (a)	145. (a)
146. (a)	147. (b)	148. (b)	149. (a)	150. (b)
151. (c)	152. (c)			

12. Electrothermism and Electroluminance

1. (b)	2. (d)	3. (b)	4. (b)	5. (b)
6. (b)	7. (c)	8. (b)	9. (a)	10. (b)
11. (b)	12. (b)	13. (a)	14. (a)	15. (b)
16. (c)	17. (a)	18. (c)	19. (b)	20. (c)
21. (a)	22. (a)	23. (a)	24. (a)	25. (c)
26. (a)	27. (b)	28. (d)	29. (b)	30. (c)
31. (c)	32. (c)	33. (c)	34. (a)	35. (c)
36. (d)	37. (d)	38. (d)	39. (b)	40. (b)
41. (b)	42. (c)	43. (a)	44. (b)	45. (a)
46. (b)	47. (b)	48. (a)	49. (d)	50. (c)
51. (c)				